SUPPLEMENT TO THE
PROCEEDINGS OF THE SEMINAR FOR ARABIAN STUDIES
VOLUME 40

THE DEVELOPMENT OF ARABIC AS A WRITTEN LANGUAGE

Papers from the Special Session of the Seminar for Arabian Studies held on 24 July, 2009

edited by
M.C.A. Macdonald

SEMINAR FOR ARABIAN STUDIES

ARCHAEOPRESS
OXFORD

2010

This Supplement is available either as a set with volume 40 of the *Proceedings* or separately from Archaeopress, Gordon House, 276 Banbury Road, Oxford OX2 7ED, UK.
Tel/Fax +44-(0)1865-311914.
e-mail bar@archaeopress.com
http://www.archaeopress.com
For more information about the *Proceedings*, see the Seminar's website: http://www.arabianseminar.org.uk

Seminar for Arabian Studies
c/o the Department of the Middle East, The British Museum
London, WC1B 3DG, United Kingdom
e-mail seminar.arab@durham.ac.uk

Typesetting, Layout and Production: Dr. David Milson

This Supplement is produced in the Times Semitic font, which was designed by Paul Bibire for the Seminar for Arabian Studies.

ISSN 0308-8421
ISBN 978-1-905739-34-9

The Steering Committee of the Seminar for Arabian Studies is most grateful to the

MBI Al Jaber Foundation

for its continued generosity in making a substantial grant

towards the running costs of the Seminar and the editorial expenses of producing the

Proceedings of the Seminar for Arabian Studies

and this Supplement.

The Steering Committee would also like to thank the following for generously

contributing to the costs of the Special Session on

The development of Arabic as a written language

at the 2009 Seminar, the papers from which are published here:

The Leigh Douglas Memorial Fund

The Oxford Centre for Islamic Studies

The Seven Pillars of Wisdom Trust

Contents

Preface

New research on Arabia in the humanities is regularly presented at the Seminar for Arabian Studies, a three-day international conference held annually at the British Museum.[1] Naturally, the programme is always extremely varied, with papers covering a timescale from the Palaeolithic to the end of the Ottoman empire, and subjects ranging from languages, inscriptions, and archaeology, to ethnography, sociology, history, art history, architecture, literature, and much more.

This tradition of variety of content is one of the Seminar's strengths and the range has continuously expanded over the forty years of its existence. At the same time, the Seminar's Steering Committee has recently felt that the conference should also provide a forum for more detailed explorations of specific topics on which new evidence or important new research had recently appeared. It therefore decided to initiate occasional "Special Sessions" dealing in detail with a specific subject. In 2007, the first of these was on *The Palaeolithic of Arabia* and was judged to be a great success.[2]

A second Special Session, this time on *The development of Arabic as a written language*, was held at the 2009 Seminar on Friday, 24th July. It was organized by Robert Hoyland, Venetia Porter and myself and consisted of a very full morning in which we heard an introduction and nine papers; followed, in the afternoon, by more papers and a lively and enjoyable workshop in which points raised in the morning were discussed and inscriptions, manuscripts and other items relevant to the Session's theme were presented.[3]

On the previous evening, I had had the honour of giving the MBI Al Jaber Foundation public lecture at the British Museum. This annual lecture generously funded, by the Seminar's major benefactor, takes place each year on the first evening of the Seminar. I chose as my topic *Ancient Arabia and the written word* as an attempt to present the background to the following day's Special Session. An edited version of this lecture, into which I have incorporated the paper I gave in the Session, will be found in this volume.

In contrast to the Seminar itself, the speakers in this Special Session were all invited, and the general areas on which they would speak suggested to them in order to give the Session an internal coherence. We are most grateful to all the speakers for agreeing to this structure and for the extremely stimulating content of their papers. We would also like to thank Professor Janet Watson and Professor Hugh Kennedy for chairing the morning session and the Workshop so admirably.

The Special Session was funded by the Seminar, and by the MBI Al Jaber Foundation, the Leigh Douglas Memorial Fund, the Oxford Centre for Islamic Studies, and the Seven Pillars of Wisdom Trust. To all these bodies we express our deep gratitude and we hope that they will find an appropriate acknowledgement of their generosity in the content of this volume.

Because the papers invited for the Special Session dealt with different aspects of a single specific topic, the Steering Committee felt that it would be best to publish them together in a separate book. It has therefore created for this purpose an occasional series of Supplements to particular volumes of the *Proceedings of the Seminar for Arabian Studies*, which will be available, either as a set with the relevant volume of *PSAS*, or separately. The present book is thus a Supplement to the *Proceedings of the Seminar for Arabian Studies* volume 40, 2010.

Robert Hoyland helped with the translation into English of Christian Robin's Introduction and Pierre Larcher's paper into English, and kindly assisted with the editing of ᶜAlī Al-Ghabbān's paper, for all of which I am most grateful to him. I would also like to thank Helen Knox, Rajka Makjanić, David Milson, and Paul and Janet Starkey for their considerable help in the preparation of this Supplement. However, any editorial faults are, of course, my responsibility alone.

[1] See www.arabianseminar.org.uk

[2] See *Proceedings of the Seminar for Arabian Studies* 38, 2008: 1–69.

[3] See the list of the papers presented, at the end of this volume.

The style and transliteration systems in this Supplement match those of *PSAS*, with the exception that, in discussions of letter forms, single letter, rather than two-letter, transliterations are used, thus: ḏ (not *dh*), ġ (not *gh*), ẖ (not *kh*), š (not *sh*), ṯ (not *th*).

I have provided a general subject index and an epigraphic index listing the numerous inscriptions, manuscripts, papyri discussed. I hope that these will be helpful to the reader as guides to the various treatments of the numerous topics discussed in this volume. The Seminar's traditional very tight schedule, by which the papers are published exactly one year after they were given, meant that there was very little time between the final setting of the volume and the date when it had to go to the printers. Thus both indexes had to be made at great speed. I therefore apologise in advance for the inevitable errors that will have crept in and hope that they will not reduce the usefulness of the indexes too much.

New discoveries and increasingly sophisticated studies are providing more and more evidence for the interplay of writing and of oral transmission in late antique Arabia. This volume of papers is presented in the hope that it will further our understanding of this fascinating subject and stimulate new explorations and research.

M.C.A. Macdonald
Oxford, July 2009

M.C.A. Macdonald (ed.), *The development of Arabic as a written language.* (Supplement to the Proceedings of the Seminar for Arabian Studies 40). Oxford: Archaeopress, 2010, pp. 1–4.

Introduction — The development of Arabic as a written language

Christian Julien Robin

Summary
Any discussion of how Arabic came to be used as a written language is based on two types of sources. One consists of the traditions concerning their past which the Arabs collected after the rise of Islam, and the other of original texts from the pre-Islamic era which have survived the hazards of history. The second category is constantly being augmented, so a regular re-examination of the evidence will always open new perspectives on the subject.

Keywords: Arabic language, Arabic script, poetry, inscriptions

It is a good idea to devote a Special Session to the subject of the development of Arabic as a written language. The Arabic language, as we know, is a key part of the national identity of the Arabs. It is also central to the religious identity of Muslims, for whom it is the language of the Revelation, and thus the language of heaven. There are of course other opinions on the language of heaven: for the Jews it is Hebrew and for the Christians of Syria it is obviously Syriac.

Academic interest in the Arabic language is far from new. Outside the Arabo-Islamic world, it was already present in Europe in the sixteenth century, while the more specific question of the development of the Arabic language in ancient times as a means of communication and composition, had already engendered numerous studies and heated debates 150 years ago.

In fact, our sources are exceptionally varied and abundant and consist of a vast corpus of reports and anecdotes gathered by the Arabs of the early centuries of Islam to illustrate their past. Although this "Arabo-Islamic tradition" contains various opinions on the creation of the Arabic script, only one would seem to have any historical value. This places its invention in the valley of the Euphrates, whence it spread first to al-Ḥīrah, and then to the Arabian Peninsula, and particularly to Mecca (Robin 2006: 327). This tradition also informs us that writing was hardly practised at all at the time of Muḥammad.[1]

For the period immediately before Islam, we hear of only two centres where writing was taught. One was at Yathrib: the Jewish *bayt al-midrās* of Māsikah/Qaynuqaᶜ at al-Quff (Lecker 1997: 263–264; al-Iṣfahānī 1994, xii: 267 = 12/6). The other was at Najrān, where it was said that a bishop was "the school master", *ṣāḥib madārisi-him* (Ibn Hishām n.d.: 579; Guillaume 1955: 271).

In addition to these relatively abundant narrative sources, we also have an equally important number of everyday documents. On the one hand, there are papyri from Egypt and Palestine. These are often dated: the earliest, found at Ahnās, being from 22/643 while others, from Naṣṭān (Nessana in Greek) in Palestine date from the early Umayyad period. On the other hand, there are inscriptions and coins (Hoyland 1997: 687–703).

Since the study of the development of the Arabic language is by no means a recent phenomenon, one can legitimately ask whether our Session can provide any new insights.

Surprisingly, the answer is a definite "yes". The main reason is that our predecessors were not very interested in everyday documents. They based their research on the Arabic narrative sources, i.e. on what the Arabs said about their own history. A good example is provided by the papyri of Naṣṭān, which I have just mentioned. They are the earliest documents of the Arab fiscal administration, and are composed in Arabic and Greek. Yet the huge *Encyclopaedia of Islam* has not a single reference to them. An initial field of research is thus open to us, namely the comparison of the narrative sources with the information from everyday documents.

This comparison is ongoing, and this brings me to the second interesting aspect of this Session. Every year, more documents come to light, most frequently in the Arabian Peninsula, and particularly in Saudi Arabia. I would like to present four examples of recent progress

[1] However, see Ghabbān (this volume). [Ed.]

in this sphere. For the first, I shall take the date of the oldest documents in the Arabic language (treating as "Arabic" documents using the definite article *al-*). In the mid-twentieth century — and for this I am using the synthesis produced by Régis Blachère (1947, i: 4–5) — the oldest documents in the Arabic language dated to the sixth century AD. These were two Christian inscriptions from Syria found at Zebed (in the region of Aleppo) and at Ḥarrān (in the Lajāʾ, south of Damascus). Today, we can take this date back more than ten centuries (Robin 2003). Thus, the period of putting things in writing and of codification began much earlier than we thought. Nevertheless, another question comes to mind: why was this written Arabic restricted to peripheral uses in Arabia up to the sixth century AD?

Other important new discoveries can be mentioned. One concerns the formation of the Arabic script. Ten years ago a number of graffiti were discovered in Saudi Arabia, which constitute the "missing link" between the Late Nabataean and the Arabic scripts. These discoveries support the hypothesis that the Arabic alphabet was derived from the Nabataean.

Another discovery is that of three rhymed poems in South Arabia, dating from the first to the fourth centuries AD (Stein 2008). These reopen discussion on the pre-Islamic Arabic poetry and the question of whether there was a specific poetic language (sometimes called a *koinè*) and, more generally, of the language which served as a substrate to Quranic Arabic.

Finally, I would like to mention three recently published Arabic inscriptions, dating to the twenties of the hijrah:

From the region of Yanbuʿ

Kawatoko 2005: 52, pl. 8/13a, from al-Muthallath, apparently in the region of Yanbuʿ

1. *ktb slmh*
2. *thlth w-ʿshryn*

1. *kataba Salmah*
2. *thalāth wa-ʿishrīn*

From the region of Madāʾin Ṣāliḥ

Ghabbān 2008, from Qāʿ al-Muʿtadil, 17 km south of Madāʾin Ṣāliḥ/al-Ḥijr

1. *b-sm ʾllh*
2. *ʾnʾ zhyr ktbt zmn twfy <h> ʿmr snh ʾrbʿ*
3. *w-ʿshryn*

1. *bi-smi ʾllāh*
2. *anā Zuhayr katabtu zaman tawaffā ʿUmar sanat arbaʿ*
3. *wa-ʿishrīn*

From the region of Najrān

Kawatoko, Tokunaga & Iizuka 2005: 8–10, from al-Khushaybah, approximately 110 km north-north-east of Najrān

1. *trḥm ʾllh ʿly yzyd bn ʿbd ʾllh ʾl-slly*
2. *ktb fy ǧmdy mn snh sbʿ*
3. *w-ʿshryn*

1. *taraḥḥama Allāh ʿalā Yazīd b. ʿAbd Allāh al-Salūlī*
2. *kataba fī jumādā min sanat sabʿ*
3. *wa-ʿishrīn*

Note
— the absence of *alif* representing [a:] in medial position
— *sanat* spelt with *tāʾ marbūṭah.*

It is interesting that these three texts, all found in Saudi Arabia, come from different regions. This implies that the use of the Arabic language and script must have been spread throughout the whole of Arabia shortly after the Islamic conquest. They also confirm that the use of the era of the hijrah, created in 17/638 on the iniative of the Caliph ʿUmar (13–23/634–644), must have come into use immediately in the Muslim empire.

Thus, recent discoveries are reviving the question of the development of Arabic as a written language. This deserves re-examination because we now know so much more about the linguistic situation of Arabia in antiquity and on the eve of Islam, as will be apparent in diverse ways in the papers of this Session.

References

Blachère R.

1947. *Le Coran. Traduction selon un essai de reclassement des sourates.* (3 volumes). (Islam d'hier et d'aujourd'hui, 3). Paris: Maisonneuve.

Ghabbān ᶜA.I.

2008. The inscription of Zuhayr, the oldest Islamic inscription (24 AH/AD 644–645), the rise of the Arabic script and the nature of the early Islamic state. (Translation and concluding remarks by R.G. Hoyland.) *Arabian Archaeology and Epigraphy* 19: 209–236.

(this volume). The evolution of the Arabic script in the period of the Prophet Muḥammad and the Orthodox Caliphs in the light of new inscriptions discovered in the kingdom of Saudi Arabia. Pages 89–102 in M.C.A. Macdonald (ed.), *The development of Arabic as a written language*. (Supplement to Proceedings of the Seminar for Arabian Studies 40). Oxford: Archaeopress.

Hoyland R.G.

1997. *Seeing Islam as Others saw it. A Survey and Evaluation of Christian, Jewish and Zoroastrian Writings on Early Islam* (Studies in Late Antiquity and Early Islam, 13). Princeton, NJ: Darwin Press.

Ibn Hishām, Abū Muḥammad ᶜAbd al-Malik/ed. M. al-Saqqā, I. al-Abyārī & ᶜA. Shalabī.

[n.d.]. *al-Sīrah al-nabawiyyah li-Ibn Hishām*. (2 volumes). (Turāth al-Islām). Beirut: Dār al-Maᶜrifah.

Ibn Hishām, Abū Muḥammad ᶜAbd al-Malik/trans. A. Guillaume.

1955. *The Life of Muhammad*. A Translation of Ibn Isḥāq's *Sīrat Rasūl Allāh*, with Introduction and Notes. Oxford: Oxford University Press.

al-Iṣfahānī, Abū ʾl-Faraj ᶜAlī b. al-Ḥusayn/ed. Maktab taḥqīq dār iḥyāʾ al-turāth al-ᶜarabī, Bayrūt.

1994. *Kitāb al-Aghānī*. (24 volumes). Beirut: Dār iḥyāʾ al-turāth al-ᶜarabī.

Kawatoko M.

2005. Archaeological Survey of Najran and Madinah 2002. *Atlal* 18: 46–59 [English section], 141–151 [Arabic section], pls 8/1–14.

Kawatoko M., Tokunaga R. & Iizuka M.

2005. *Ancient and Islamic Rock Inscriptions of Southwest Saudi Arabia*. i. *Wādī Khushayba*. Tokyo: Middle Eastern Culture Center in Japan and Research Institute for Languages and Cultures of Asia and Africa, Tokyo University of Foreign Studies.

Lecker M.

1997. Zayd b. Thābit, "a Jew with two sidelocks": Judaism and literacy in Pre-Islamic Medina (Yathrib). *Journal of Near Eastern Studies* 56: 259–273. [Reprinted as Lecker 1998 III].

1998. *Jews and Arabs in Pre- and Early Islamic Arabia*. (Variorum Collected Studies Series, CS 639). Aldershot: Ashgate.

Robin C.J.

2003. Review of A. Sima, *Die lihyanischen Inschriften von al-ᶜUḏayb (Saudi-Arabien)* (Rahden/Westf., 1999). *Bibliotheca Orientalis* 60: cols 773–778.

2006. La réforme de l'écriture arabe à l'époque du califat médinois. *Mélanges de l'Université Saint-Joseph* 59: 319–364.

Stein P.

2008. The "Himyaritic" language in pre-Islamic Yemen. A critical re-evaluation. *Semitica et Classica* 1: 203–212.

Author's address

Christian Julien Robin, Villa Entrebois, Chemin de la Tour de César, Plateau de Beauregard, Les Pinchinats, 13100–Aix-en-Provence, France.

e-mail christian.robin@ivry.cnrs.fr

M.C.A. Macdonald (ed.), *The development of Arabic as a written language.* (Supplement to the Proceedings of the Seminar for Arabian Studies 40). Oxford: Archaeopress, 2010, pp. 5–28.

Ancient Arabia and the written word

M.C.A. Macdonald

Summary

From at least the early first millennium BC, the western two-thirds of Arabia saw the flowering of a large number of literate cultures in both the north and the south, using a family of alphabets unique to Arabia. This happened not only in the settled areas, but among the nomads who, however, used writing purely as a pastime. These scripts died out in the north by about the third century AD and in the south by the end of the sixth. Among the written languages used in western Arabia, Old Arabic is conspicuous by its absence and seems only to have been transcribed on very rare occasions, using a variety of scripts. The Nabataeans used Aramaic as their written language and brought their version of the Aramaic script to Arabia in the first century BC. In late antiquity, the Nabataean Aramaic script gradually ceased to be employed to write Aramaic and came to be used for Arabic, which thus at last came to be a habitually written language. However, writing appears to have been used only for notes, business documents, treaties, letters, etc., not for culturally important texts, which continued to be passed on orally well into the early Islamic period.

Keywords: literacy, writing, Arabia, Old Arabic, alphabetic scripts, nomads

In about 800 BC, the regent of the Hittite city of Carchemish set up an inscription. He was a eunuch of the palace and had been charged with ruling the city during the minority of the sons of the late king, Astiruwas (Hawkins 2000: 78).[1] The regent was called Yariris, and in his inscription he listed his achievements and skills, among which he claimed to know twelve languages and at least four scripts.[2] The latter were: "the script of the city", i.e. Carchemish itself (hieroglyphic Luwian), the script of Tyre (i.e. the Phoenico-Aramaic alphabet), the script of Assyria (i.e. cuneiform), and the *ta-i-ma-ni-ti* script. The last almost certainly refers to the script used in the Arabian oasis of Taymāʾ, possibly as a representative of the alphabets of Arabia in general.[3]

The four scripts listed neatly symbolize a world with Carchemish at its centre, Phoenicia to the west, Assyria to the east, and Taymāʾ to the south. It also represents — though this must have been unconscious — the major types of writing system in the ancient Near East: hieroglyphic, cuneiform, and the two branches of the alphabet.

For while the *idea* of the alphabet was invented only once, probably in Egypt[4] sometime in the second millennium BC, the original alphabet seems to have split into two traditions at an early stage, and these appear to have developed in parallel. In the Levant there was the Phoenico-Aramaic branch, from which are descended all but one of the traditional alphabets used throughout the world today. The other branch was the South Semitic script family, which was used exclusively in ancient Arabia and its immediate environs, and is today represented only by the writing system used in Ethiopia for Gəʿəz, Amharic, etc. Thus, Arabia was unique in the ancient world in having its own branch of the alphabet and some of its inhabitants used it with great enthusiasm.

However, the available evidence points to an unexplained but very marked difference between the western two-thirds of the Peninsula and the eastern one third (Macdonald 2009*a* III: 38–41). In the west, the writings of the ancients are everywhere to be seen and we have evidence of the use of numerous languages,

[1] This is the adapted text of the MBI Al-Jaber Foundation annual lecture given at the British Museum during the 2009 Seminar for Arabian Studies. Since it was a public lecture, attended by both experts in the subject participating in the Seminar and members of the public who came to it with no previous experience, it inevitably contains information that is well known to some but new to others. It was intended as an introduction to the Seminar for Arabian Studies' Special Session on *The development of Arabic as a written language*, of which this volume is the publication, and my brief paper in that Session has been incorporated into this one.

[2] Hawkins 2000: 131, Inscription II.24 Karkamiš A15b, ll. 19–20. The beginning of the list of scripts in l. 19 is lost, so there may have been more than four.

[3] See the discussions in Hawkins 2000: 133, note on l. 19; and more recently in Macdonald 2009*a*, Addenda: 15–16.

[4] See recently Sass 2008 and references there.

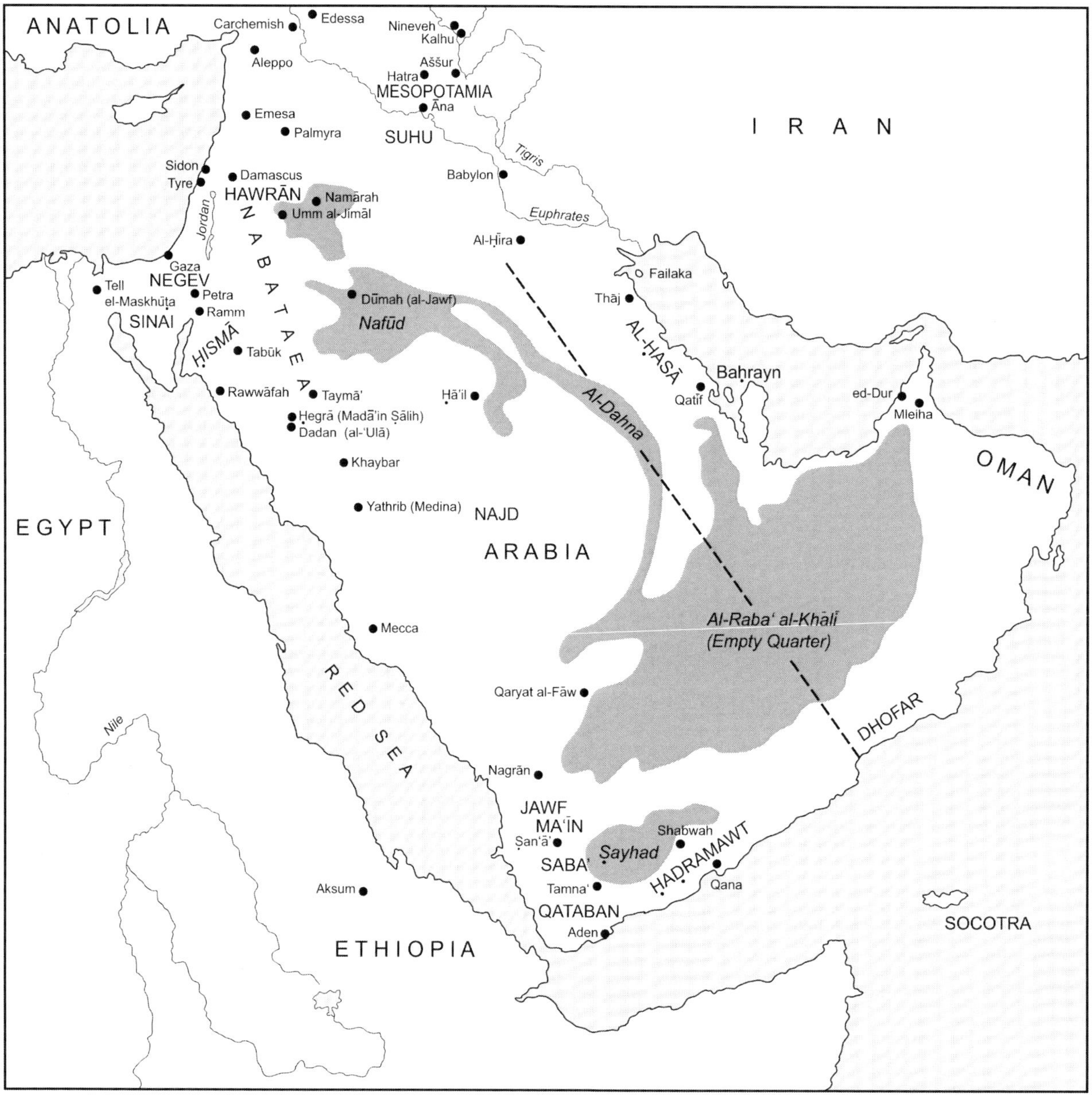

FIGURE 1. *A map of the ancient Near East showing the places and the rough east–west division mentioned in this paper. (By kind permission of Equinox Publishing Ltd).*

dialects, and scripts. By contrast, in the eastern third of the Peninsula, evidence of writing is extremely rare. Moreover, whereas in the west, several settled areas developed their own forms of the script and the nomads developed several others, in the east we know so far of only one indigenous script, that of undated and undeciphered dipinti and graffiti in Dhofar, southern Oman (al-Shahri 1991; 1994). All the other inscriptions in eastern Arabia — and there are fewer than 180 in all — are in imported scripts: Akkadian cuneiform, Aramaic, Greek, and Ancient South Arabian (see below). We do not know the reasons for this curious difference between the two sides of the Peninsula: whether it represents a difference in the levels or uses of literacy in the two regions in antiquity, or simply the marked disparity in the availability of durable writing materials.

* * *

As I hope to have shown elsewhere (Macdonald 2009*a* I), a society can be literate, in the sense that its political, administrative, religious, and sometimes commercial functions rely on writing, even when the majority of the population is unable to read or write. This was the case in mediaeval Europe for instance. On the other hand, there are some societies in which word of mouth and memory perform all the functions of communication and record for which we use writing. I would call such societies and their members "non-literate", and reserve the term "illiterate" for those who cannot read and/or write within a literate society (2009*a* I: 49–50). Just as there are often many illiterates in literate societies there can also be many people who can read or write in a non-literate society, without it affecting their continued use of memory and oral communication in their daily lives. The Tuareg nomads of north-west Africa are an excellent modern example of this. They speak Berber dialects and live in non-literate societies which function on memory and oral communication, and yet they have their own writing system, the Tifinagh, which they use purely for amusement: playing games, carving graffiti, writing coded love letters, etc. If they need to write for practical purposes, they will employ a scribe (or find a relative who has been to school) to write in Arabic or French, even if they are writing to another Tuareg who will then have to ask someone else to read the letter and translate it for him. The Tuareg have an extremely rich oral literature in which writing, even in their own script, plays no part. Culture is quintessentially oral;[5] writing in their own script is for fun; and for practical, non-cultural, activities they use writing by proxy in a foreign language and script. A very similar situation seems to have existed in ancient Arabia, and it is worth bearing this in mind as we approach the many and varied uses of literacy there.

* * *

The best-known examples of literate societies in the Peninsula are the kingdoms of ancient South Arabia. The Ancient South Arabian alphabet is known in two different forms: the *musnad* and the *zabūr*. The *musnad* was used for some 1500 years from the early first millennium BC to the sixth century AD (see Drewes *et al.* forthcoming), during which time huge numbers of public inscriptions were carved on rock faces, stelae, gravestones, and objects such as incense burners and altars, as well as being carved on, or cast in, bronze.

But what do these thousands of inscriptions tell us about the extent and nature of literacy in these kingdoms? Carving monumental inscriptions on stone, or casting them in bronze, is very skilled work, so the nominal "authors" of these inscriptions would actually have *commissioned* them, and need not themselves have been able to write, or even able to read the finished product. Even the so-called penitential or confession inscriptions, which sometimes acknowledge very intimate sins, are formulaic in their structure and do not suggest at all that they contain the penitent's own words. It seems virtually certain that the text of these inscriptions would have been composed and written out by a temple scribe and then transferred to stone by the mason. In other inscriptions, such as those commemorating the construction of a building or irrigation system, celebrating the achievements of a ruler, or setting out rules and regulations, etc. the wording would have been dictated to the scribe who, once again, would have written out the text for the mason to copy. This means that in the process of creating these inscriptions, only one person — the one who is the least visible to us, i.e. the scribe — need have been able to write.

One should also remember that public inscriptions are often intended more as symbols than as channels of communication. In most cases in antiquity, if it was necessary to promulgate the text of the inscription, it was distributed on parchment or papyrus and/or was proclaimed. Moreover, in antiquity, as in the Middle Ages, silent reading was rare enough to be remarked on, and reading aloud was the norm, so it only required one literate person to read an inscription for all within earshot to get the message.[6] For the most part, however, I suspect the inscriptions themselves remained symbols of authority or commemoration with no requirement, or even expectation, that they would be read, a conclusion that, I am happy to say, has been reached independently by Peter Stein (personal communication). I would therefore suggest that the existence of large numbers of public inscriptions is not *of itself* an indication of widespread literacy in a society, but see below.

Until forty years ago, the *musnad* was the only known form of the Ancient South Arabian alphabet. But since the early 1970s thousands of texts have come to light in another version of the script, the *zabūr*. It developed from the *musnad* early in the first millennium BC and then the two versions evolved in parallel (see Ryckmans 2001). It was used, not for public inscriptions but for everyday documents such as contracts, letters, schedules, lists, etc. These were incised on palm-leaf stalks and sticks, where the outer skin or bark was peeled off when they

[5] See the brilliant study by Galand-Pernet (1998).

[6] For a more detailed discussion see Macdonald 2009*a* I: 99, and n. 61.

were freshly cut, revealing a relatively soft surface on which texts could be incised with a sharp blade. However, it should be borne in mind that, with the exception of twenty-two examples from the site of Raybūn in Ḥaḍramawt (Frantsouzoff 1999), all the sticks known so far appear to come from a single site — a huge archive which must have been used over a period of 1500 years near the ancient town of Nashshān (modern al-Sawdāʾ), in northern Yemen (Stein 2005*a*: 184).

Who used the *zabūr*? One might assume that the thousands of sticks that have survived imply widespread literacy. However, there are a number of factors which may suggest that this was not so. For instance, in the correspondence carved on these sticks, while the recipient is addressed as "you", the sender appears as "he" or "she", rather than "I", which suggests that an intermediary, such as a scribe, was actually *writing* the letter. Indeed, after a meticulous study of these documents over many years, Peter Stein has suggested that when someone in ancient South Arabia wanted something written, he or she would go to a scribal centre (with its archive), where the document would be written for them, and possibly a copy retained (Stein 2005*b*: 148–150). So even the existence of a large number of informal and personal documents does not necessarily indicate widespread literacy.

However, there is one class of texts, which strongly suggests that literacy was more widespread in ancient South Arabia, than would appear from the above. For, as well as thousands of public inscriptions and documents on sticks, there is an abundance of graffiti. These are found all over what is now Yemen and in the deserts to the north, between Najrān and Qaryat al-Fāw (Fig. 1). As one would expect, they are carved in the *musnad*, i.e. the script of public inscriptions, rather than the *zabūr*, since in most cultures in which there are formal and informal versions of the script, graffiti tend to be in the formal version, like public inscriptions. Thus, in the Greek, Roman, and Cyrillic alphabets, capital letters are normally used for graffiti (even those in spray paint),[7] and I would suggest that this is why unpointed angular Kufic was used for several centuries in Arabic graffiti even though it was hardly, if ever, used in everyday documents.

Now, most people, even if they are literate, get little chance to write the version of the script used for inscriptions in their society. There is no exact equivalent in the West because we use both capitals and lower-case letters in our daily writing and so have no formal version of the script, although inscriptions and graffiti tend to be only in capitals. However, most of us, for instance, have a reading knowledge of the letter forms of typefaces, but very few of us have any practice in reproducing them by hand, although if we have a relatively good visual memory we could make a reasonably successful attempt. The same would be true of people in ancient South Arabia carving graffiti in the *musnad*, which is probably why we find some attempts that are not entirely successful. There is a famous remark in Petronius' *Satyrica* (LVIII.7) where a Roman freedman says that although he had had no formal education, *lapidarias litteras scio* "I know the letters used in inscriptions".[8] This gives us an insight into how reading literacy could spread informally in a society in which the majority of people use their memories a great deal more than we do, partly because they cannot use writing as a substitute. It does not take very long or an enormous effort, to learn to read an alphabetic script, particularly if one is learning only one version (for instance the capitals used in Roman inscriptions, or the *musnad* in their Ancient South Arabian equivalents). The problem comes when one tries to transfer this *reading* knowledge of the letters to writing them — or, in the case of most ancient graffiti, carving them — if one has had no training and very little practice in writing.

Graffiti are, by definition, the work of individuals and it is highly unlikely that a professional stonemason would be employed to carve a graffito for someone. What would be the point?[9] Indeed, I would say that once an inscription is commissioned it ceases to be a graffito. There is not

[7] Robert Hoyland has pointed out to me (personal communication) that young children have traditionally used capitals in writing birthday cards or labelling their drawings and that this is presumably not out of a desire for formality. However, I would suggest that they do so simply because they were taught the capital letters first and only later the lower-case forms. As teaching methods have changed I have noticed that cards and notes from some young children now tend to be all in lower case, with no capitals.

[8] On this see Macdonald 2009*a* I: 77, n. 91.

[9] In the years shortly after the Safaitic script was finally deciphered in 1901, Enno Littmann, among others, suggested that the fact that most Safaitic inscriptions begin with the preposition *l* (the so-called *lām auctoris*) and were expressed in the third person singular, suggested that they were written by scribes (1904: 111; see also 1940: 98–99; 1943: viii). However, given the vast numbers of these graffiti this would have been a logistical impossibility and the idea of employing a scribe to carve a graffito is anyway incongruous. On the *lām auctoris* see Macdonald 2006: 294–295. The use of the first or third person in a graffito is surely simply dictated by the introductory formula employed. If it begins "I (am) so-and-so . . ." it is natural for it to continue in the first person. If it begins "By so-and-so . . ." (like the vast majority of Safaitic graffiti) it is natural for it to continue in the third person. This is quite different from the case of the ancient South Arabian correspondence incised in the *zabūr* on sticks, where we have the equivalent of indirect speech ("he asks you"), as explained above.

a simple dichotomy between formal ("monumental") inscriptions and graffiti. There are plenty of other kinds of carved texts that fit into neither category (e.g. at random, *me fecit* or magic inscriptions on objects; exhortations such as "Vote for X"; *cave canem*; or announcements of entertainments, closure of public buildings, etc.).

Thus, if, as I am suggesting, a graffito is by definition carved or written in the author's own hand, and the very large numbers of graffiti in South Arabia are in a script (i.e. the *musnad*) which would not normally have been used in day-to-day writing, this suggests that there may well have been a fairly widespread *reading* knowledge (at least) of the formal *musnad* script in the general population.

While it is clear, therefore, that the ancient South Arabian kingdoms were literate societies in the sense that they relied on the written word for important functions, we do not have sufficient evidence to know how widely even reading-literacy, let alone the ability to write, was spread throughout the population. However, I would suggest that, at least at some periods, quite considerable numbers of people in these societies must have been able to read the *musnad* script used in public inscriptions and managed to convert this reading knowledge to a writing knowledge in order to carve graffiti. We cannot deduce from this, however, that they practised writing in other circumstances. Normal life outside the palace chancellery, the temple, and probably the merchant's office, almost certainly did not require ordinary people to write, and those who did write would presumably have used the *zabūr* in their daily life, not the *musnad*. Thus, ironically, the vast numbers of inscriptions produced in ancient South Arabia do not necessarily imply a very high degree of literacy in the population. For that we have to go north and to a very unexpected group. But first we should examine what we know of the social, commercial, and cultural situation in north and central Arabia.

* * *

Here, a different pattern emerges. In the first millennium BC, frankincense was probably the most valuable commodity on the markets of the ancient Near East and the Mediterranean world. South Arabia was the only source of good-quality frankincense, which is a resin tapped from trees of the species *Boswellia sacra* in the mountains of Dhofar. But this source in the south of the Peninsula was far away from the almost insatiable markets in the north and the frankincense had to be brought from source to market by camel caravans travelling across Arabia. This involved the nomads of north and central Arabia who provided the camels and those who looked after them, the guides and the guards, and who no doubt charged for the privilege of crossing the territories within which they migrated. It also involved the inhabitants of the settled areas, who would not only have charged tolls, but would have sold the members of the caravan food and water for themselves and their beasts, and would no doubt have set up profitable markets for the exchange of goods. All this distributed the wealth generated by the trade over large areas of Arabia drawing huge amounts of money and goods into the Peninsula from Mesopotamia, the Near East, Egypt, and the Mediterranean. In a financial sense, frankincense was the petroleum of antiquity.

But it had other effects as well. In antiquity, the great oases of north-west Arabia were cosmopolitan trading centres with links to the great kingdoms surrounding them, and even, as we have seen, as far away as Carchemish in what is now southern Turkey. I have argued elsewhere that the merchants from the oases not only sold frankincense in these areas, but may have traded between centres in the Levant and Mesopotamia in goods bought in one place and sold in another, as well as returning to Arabia with goods purchased in the north (Macdonald 2009*a* IX: 339–340). A dramatic account from the mid-eighth century BC by the governor of Suḫu and Mari on the Euphrates tells how he raided a caravan of "the people of Tema and Saba whose own country is far away" just after it had visited the city of Ḫindānu, apparently on its return journey to Arabia (ibid. pp. 338–340 and references there).[10] Although the list of the booty he took from it is damaged, it includes purple cloth, wool, precious stones, and iron, the latter a commodity which we know from another Assyrian document was sought after by the "Arabs" (e.g. Parpola 1987: 140, no. 179, lines 22–23), but no frankincense.

Each of the three major oases of north-west Arabia — Taymāʾ, Dedān, and Dūmah — developed its own form of the South Semitic alphabet. This in itself is interesting since they were geographically closer to areas using the Phoenico-Aramaic alphabets than they were to South Arabia, and at first sight one might have expected them to have adopted the Phoenico-Aramaic script from the Levant. On the other hand, they do not seem to have taken the alphabet from South Arabia either. Although we still do not understand the exact interrelationships of the various members of the South Semitic script family, it

[10] Naʾaman assumes, on no evidence that I can discover, that the "caravan was on its way north to the territory under Assyrian rule" (2008: 234).

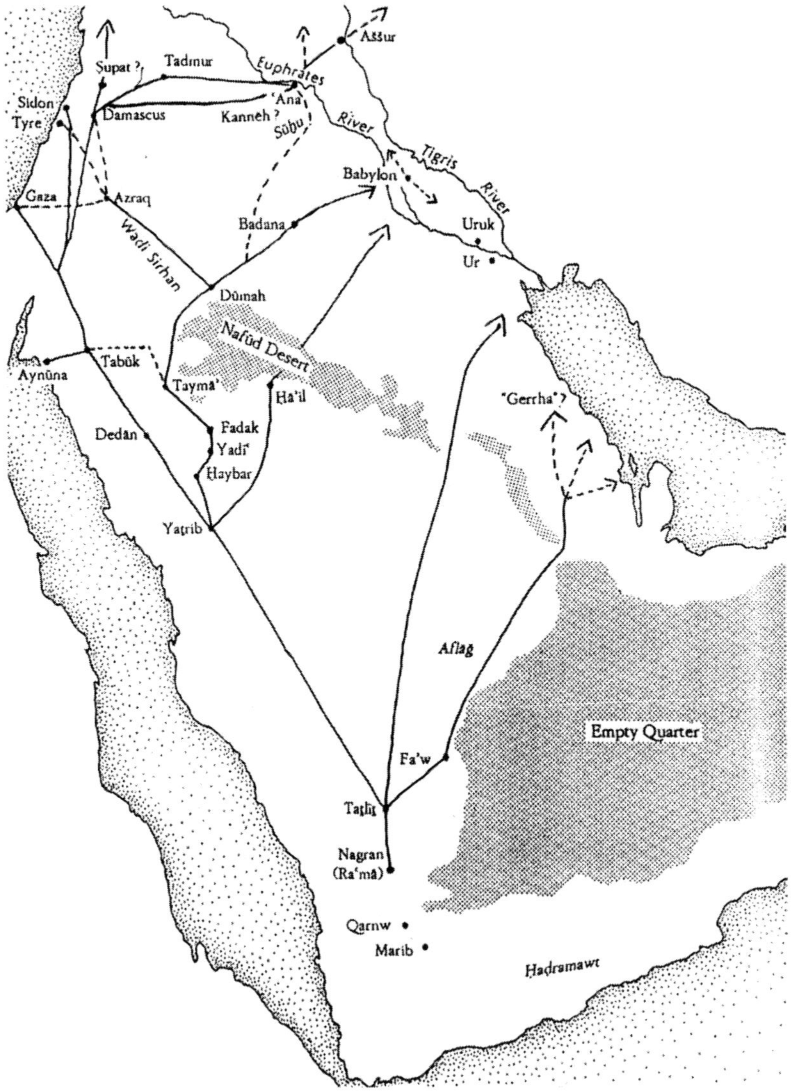

FIGURE 2. *A map showing trade routes across Arabia in the first millennium BC and the early centuries AD. (From Macdonald 2009*a *IX: 349, by kind permission of Ashgate Publishing Ltd).*

seems likely that the alphabets used in South Arabia and those used in North Arabia developed in parallel rather than one from the other.

I mentioned earlier that Yariris, the regent of Carchemish in about 800 BC claimed to be able to read the alphabet of Taymāʾ. There are several hundred inscriptions and graffiti in this script scattered around the oasis and its environs, and although none found so far mentions Carchemish, they tell us, among other things, of wars with Dedān, the great rival oasis that dominated the other major west Arabian caravan route to the north (Fig. 2 and Macdonald 2009*a* IX: 334–336). The heavy involvement of the oases and nomads of North Arabia in the frankincense trade encouraged the Assyrians to try repeatedly but unsuccessfully, to subjugate them in the eighth to seventh centuries BC, when they fought campaign after campaign against successive queens and kings "of the Arabs" (Ephʿal 1982: 81–191). In the mid-sixth century Nabonidus, the last king of Babylon, conquered six oases in north-western Arabia[11] and took

[11] These were Taymāʾ, Dedān, Fadak, Khaybar, Yadīʿ, and Yathrib (Gadd 1958: 80–85), see Fig. 2.

up residence in Taymāʾ for ten years between 552–543 BC (Gadd 1958; Beaulieu 1989: 149–185). We know this not only from Nabonidus' own inscriptions but also from graffiti in the Taymanitic alphabet, which mention his presence and that of his officials (Hayajneh 2001*a*; 2001*b*; Müller & Said 2001). It will be clear from Figure 2, that by conquering Taymāʾ, Dedān, and Yathrib (modern Medina), Nabonidus gained control of the northern parts of all the western routes of the frankincense trade (Macdonald 2009*a* IX: 334–336, 349).

There are several curious things about the script of Taymāʾ. So far, apart from a number of gravestones, all the texts carved in it are graffiti. There are no official government inscriptions, nor are there any official religious texts in this script, although a number of graffiti contain prayers and religious statements. This is the situation so far, and it should be noted that the vast majority of these graffiti are from the *environs* of Taymāʾ rather than inscriptions from the oasis itself, though this may simply be because the graffiti on desert rocks have been left undisturbed in contrast to the continuous occupation of the oasis. However, the ongoing Saudi-German excavations in the oasis may change this at any moment. Like all the alphabets of the South Semitic script family except Dadanitic, on which see below, the Taymanitic alphabet consists only of consonants, and vowels are normally not represented. Like Ancient South Arabian and the scripts used in the other North Arabian oases, Taymanitic often marks the division between words by word-dividers, usually in the form of short lines or dots.

Taymanitic graffiti can be written in any direction, but the majority are written horizontally from right to left or left to right and, when there is more than one line, in boustrophedon, an arrangement generally found in scripts which are used principally for carving, rather than for writing with ink. This is because, in a script where each letter is separate, rather than joined to the one that follows, the direction is of little consequence to the stonemason or to someone scratching or hammering a graffito on a rock face. However, if you are writing with a pen, it is difficult to cut a nib that can write equally fluently in both directions. The same may have been true of the blades used to incise texts on sticks in South Arabia, since I am informed by Peter Stein (personal communication) that, without exception, even the earliest of these from the tenth century BC, runs only from right to left, even though, some 200 years later, some inscriptions on stone in the *musnad* run boustrophedon. This suggests that boustrophedon is not simply an early stage in the development of a script, as is usually assumed, but was a conscious aesthetic choice by the designer of the inscription.[12]

One of the lasting effects of Nabonidus' sojourn in Taymāʾ, was the introduction to the oasis of Aramaic as the language and script of prestige. Over the following decades, the Taymanitic alphabet seems to have died out, or at least we do not yet have any texts datable to a later period. Instead, the only inscriptions from the later times are in Aramaic. At first, it was the Imperial Aramaic used by the bureaucracies of the Babylonian and later the Achaemenid Persian Empire.[13] It is interesting that when — probably in the Achaemenid period — the kings of Liḥyān ruled Taymāʾ, the inscriptions they set up in Taymāʾ were carved in Imperial Aramaic,[14] whereas those which the same kings left in their own oasis, Dedān, were carved in the local (South Semitic) Dadanitic script (see below). Later, probably after the fall of the Achaemenid empire and the end of the regularizing influence of its chancellery, there developed a form of the Aramaic script which seems to have been peculiar to the oasis itself.[15] Probably, in the late first century BC this was supplanted by the Nabataean version of the Aramaic script, when Taymāʾ seems to have come under Nabataean cultural influence.[16]

---

[12] It is interesting that Pirenne (1956: 97), followed by many subsequent scholars working on Ancient South Arabian inscriptions, considered that "le boustrophédon constitue une *caractéristique suffisante* pour attribuer une inscription à cette période [sc. that of her earliest "graphies", A and B]" (italics in the original), even though she recognized that other inscriptions from the same period were unidirectional. However, there are also boustrophedon inscriptions from much later, fifth–fourth centuries BC for instance, at the Barāʾn Temple in Maʾrib; see for instance Daum *et al.* 2000: 285, no. 35M; Nebes 1992: 162.

[13] See *CIS* ii 113–116; Degen 1974, nos 1–10; Cross 1986; Beyer & Livingstone 1987; 1990.

[14] These inscriptions were found by the Saudi-German excavations at Taymāʾ. See Deputy Ministry of Antiquities and Museums 2007: 31 (photograph); al-Said in press; Eichmann, Schaudig & Hausleiter 2006: 168; and Eichmann 2009: 61. Several others have since been discovered. I am most grateful to Professor Eichmann and Dr Hausleiter for inviting me to participate in the 2010 season of excavations at Tayma and to Dr Muhammad al-Najem, director of the Taymāʾ Museum, for giving me access to the inscriptions in the Museum storeroom.

[15] *CIS* ii 336 = Milik 1978.

[16] Although a number of inscriptions in Imperial Aramaic and the local form of the Aramaic script have been known for many years (see the previous two notes), the first texts in the *Nabataean* script from Taymāʾ were found in the recent Saudi-German excavations; see Eichmann 2009: 59–66. In March 2009, a Nabataean inscription dated to AD 204 was discovered during roadworks in Taymāʾ; see Al-Najem & Macdonald 2009.

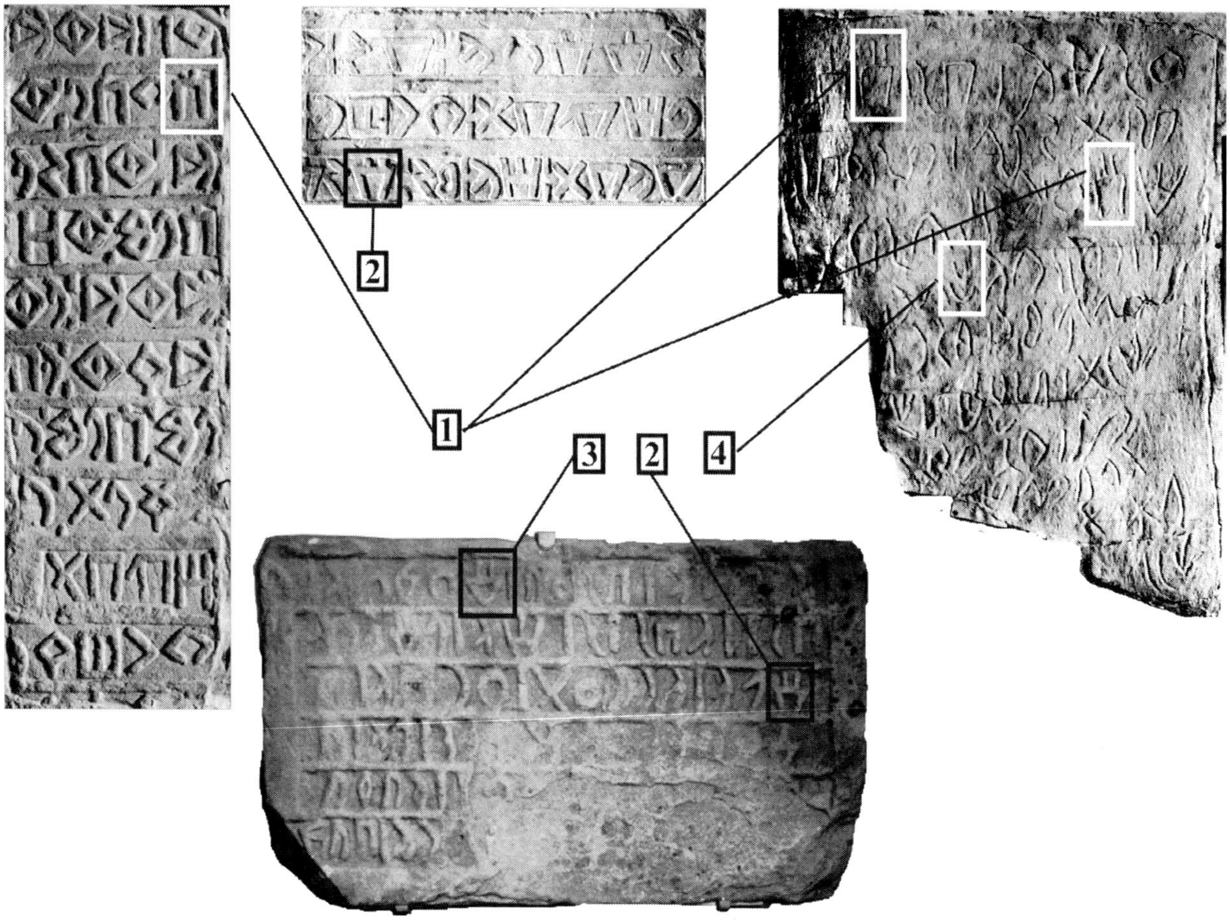

Figure 3. *Developments in the shape of Dadanitic* ʾ*. The numbers refer to the different forms of the letters as discussed in the paper. From left to right: JSLih 49, JSLih 42 (upper), Said 1999: 15–25, no. 2 (lower), and JSLih 71.*

The large oasis of Dedān (modern al-ᶜUlā) dominated the other great route to the north (Fig. 2; Macdonald 2009*a* IX: 334, 337–338, 341–343). It too developed an alphabet of its own, probably over a considerable period, though we have little or no firm dating evidence as yet.[17] At Dedān, we have a considerable number of public inscriptions, mostly carved in relief, plus several hundred graffiti, but as at Taymāʾ, no documents on perishable materials comparable to the sticks found in South Arabia. On the other hand, there are certain indications that at Dedān such documents, either written in ink or incised on soft wood, may have existed. All the inscriptions and virtually all the graffiti are written from right to left and there are no texts in boustrophedon. As I mentioned in connection with the Ancient South Arabian scripts, writing in only one direction usually develops because of the practical requirements of pens and possibly blades, and is a matter of indifference to the stonemason. The fact that we have no inscriptions in boustrophedon at Dedān therefore suggests — and, of course, it can be no more than a suggestion — that the script had been used for documents written in ink or possibly incised on wood for some time before it came to be carved on stone.

Moreover, certain letters develop forms which it is difficult to explain if the script was used only for carving on stone and which are more likely to have developed through writing with a pen. This can be seen, for instance, in the development of the shape of *alif* (see Fig. 3). In the formal version it has straight vertical "legs" (as in

[17] Caskel's attempts to create a palaeographical sequence (1954: 21–44), are based on an abuse of palaeographical and historical method; see Macdonald, forthcoming, *a* and *b*.

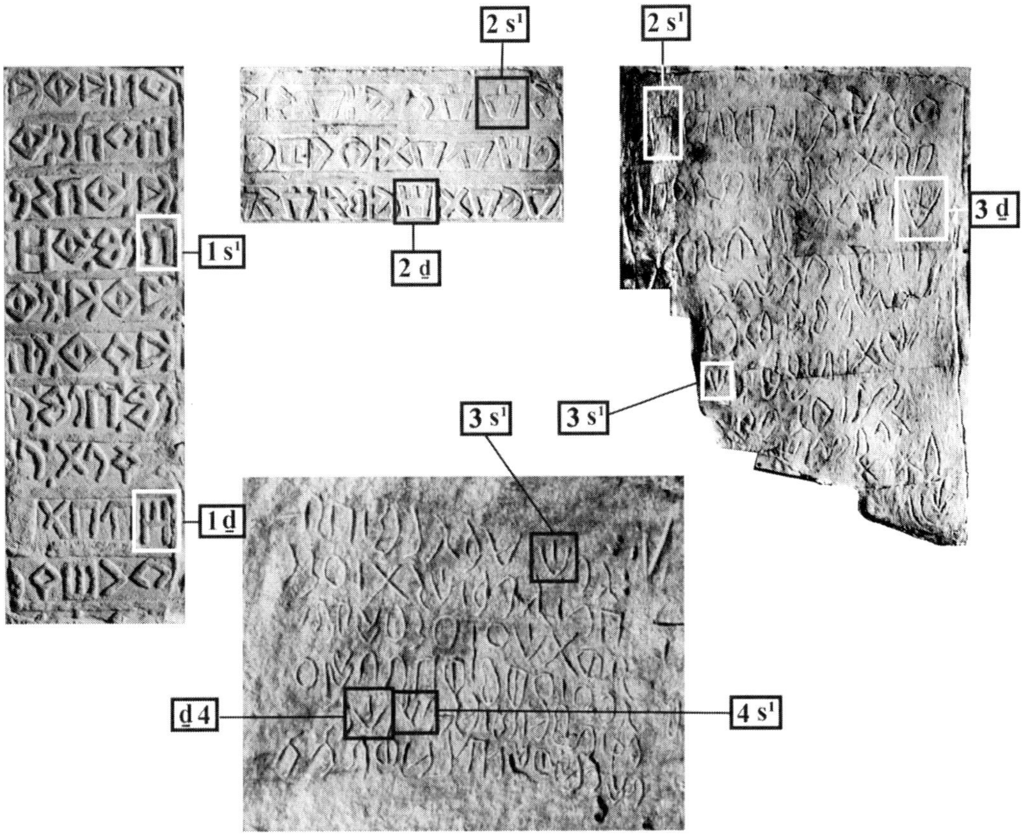

Figure 4. *Developments in the shapes of Dadanitic ḏ and s^1. From left to right: JSLih 49, JSLih 42 (upper), JSLih 70 (lower), JSLih 71.*

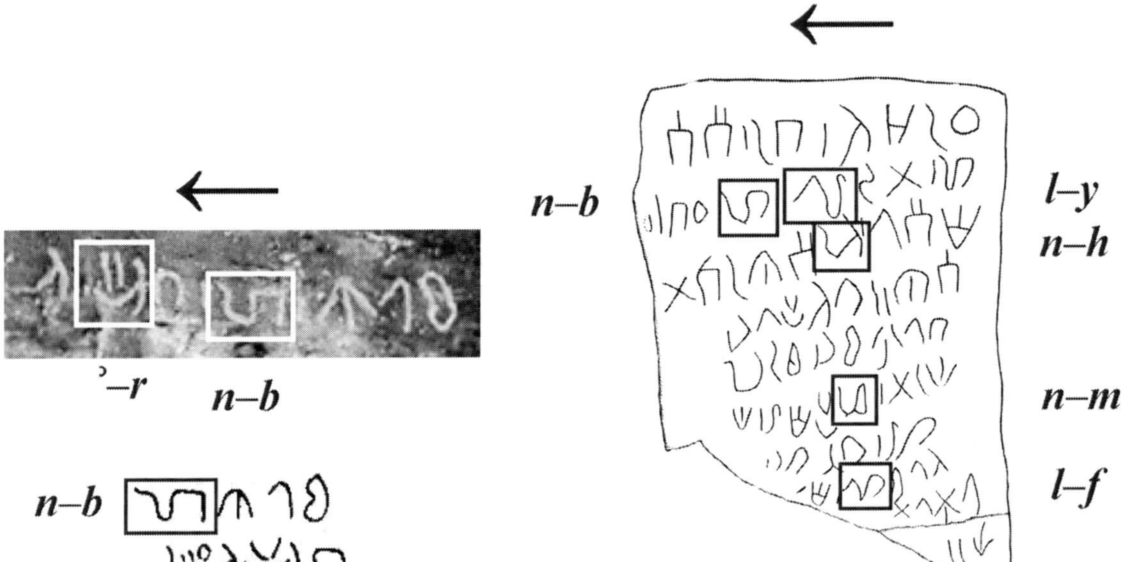

Figure 5. *Ligatures in Dadanitic inscriptions. The arrows show the direction of writing. On the left, two different graffiti by the same man, Grḥ bn Br'h, the upper is unpublished and the lower is JSLih 375. On the right JSLih 71.*

1), but it can be seen that there is a tendency for these to converge (as in 2) and even to form a triangle (as in 3) and eventually the horizontal bar disappears and it becomes two inverted chevrons (as in 4). This form is regularly found in the same text as ones with vertical or converging legs, as can be seen in the photograph on the right of Figure 3.

A similar process takes place with the form of the letter *ḏāl* and with that of s^1 (Fig. 4), two letters which, from having completely different shapes in the formal versions (as in 1), end up with almost identical informal shapes (as in 4). It is important to note that, with all these letters, the informal shapes must have evolved *in parallel* with the use of the formal ones, since we regularly find them used side by side in the same inscription. It is strange, but it appears that the stonemasons and those who employed them, considered the informal shapes to be valid alternatives to the formal ones, even within the same text.

There are also occasional examples of ligatures joining letters (Fig. 5). Ligatures only develop when writing with a pen since they increase the ease and speed of writing by removing the necessity of lifting the pen between letters. If found in a graffito, therefore, they are generally a good indication that the carver is used to writing in pen and ink.[18]

It would stand to reason that these oases, which were so heavily involved in commerce between the literate societies of South Arabia and those in Egypt, the Levant, and Mesopotamia, would use writing for record-keeping and communication. Indeed, it is difficult to see how Yariris far away in Carchemish would have known of the Taymanitic script or bothered to learn it, if it had not been used for commercial and perhaps diplomatic and legal documents. Small texts in the scripts of these oases, and other variants, have been found in Mesopotamia, Iran, Syria, and Palestine,[19] again suggesting that the merchants of the oases carried their scripts with them in their international business, something they would surely only do if they were using writing in their work.

The third great oasis of north-west Arabia was Dūmah (known as Dūmat al-Jandal in the Middle Ages and al-Jawf today). It seems also to have had its own offshoot of the South Semitic script family, but so far alas, this is known from only three graffiti (Winnett & Reed 1970: 80–81, nos. 21–23). Dūmah was in a strategic position on the south–north trade routes, since from here caravans could go north-east to Mesopotamia or due north up the Wādī Sirḥān to the Levant (Fig. 2, and Macdonald 2009*a* IX: 335–337). In the eighth and seventh centuries BC, Dūmah was the cult centre of several Arab tribes, particularly Qēdār, which the Assyrian empire tried unsuccessfully to conquer. It was an important religious centre and the Arab queens who led the resistance against the Assyrians seem also to have been priestesses (seeEphᶜal 1982: 118–123 and n. 400). The Assyrians twice carried off the images of six of the deities worshipped there[20] and it is interesting that three of these, *ᶜtrs¹m* (which appears as ^ilu^*A-tar-sa-ma-a-a-in* in the Assyrian Annals), *rḍw* (which appears as ^ilu^*Ru-ul-da-a-a-u*), and *nhy* (which appears as ^ilu^*Nu-ḫa-a-a*),[21] are invoked in the graffiti in the script of Dūmah, and in the scripts of the nomads of north and central Arabia, on which see below.

So far, I have emphasized the links between the North Arabian oases and the kingdoms to the north of them. However, the principal merchandise on which they depended came, of course, from the south, and with it came the merchants of the kingdoms of Sabaʾ and Maᶜīn. It is likely that South Arabian merchants would have used writing in their business and brought their skills with them to the north. Indeed, the members of what is assumed to have been a Minaean commercial station at Dedān left a number of public inscriptions and graffiti there and presumably wrote, or had a scribe write for them, documents on perishable materials like palm-leaf stalks, though none have been discovered there as yet. It would be very interesting to know whether the Minaeans had any influence on the writing practices of the local populations of Dedān, or vice versa.

I would suggest, then, that the picture that emerges from the settled populations of ancient west Arabia is one of literate societies in which, even if the majority of the population was illiterate, the written word was fundamental to the functioning of government, religion, and especially commerce. There must also have been a sizeable number of private citizens able to carve graffiti in the forms of the script used for public inscriptions. In

[18] On playful redundant ligatures in graffiti see Macdonald 1989; 2009*a* II: 386–387.

[19] These are the "Dispersed Oasis North Arabian" texts, on which see Macdonald 2009*a* III: 33.

[20] Sennacherib took them between 691 and 689 BC and they were returned by Esarhaddon between 681 and 676. Esarhaddon then took them away again between 673 and 669. One of them, Atarsamain, was returned by Assurbanipal in 668; see Ephᶜal 1982: 119, 125–129, 147.

[21] See Campbell Thompson 1931: 20, lines 10–11. For the identification of ^ilu^*Ru-ul-da-a-a-u* as *Rḍw*, in which *ld* is an attempt to reproduce the lateral pronunciation of /ḍ/ (cf. Spanish *alcalde* < Arabic *al-qāḍī*) see Milik 1972: 49.

South Arabia, we now have evidence of the extensive use, through scribes, of writing in day-to-day activities. In the north, we have as yet no direct evidence for the use of writing at this level, but there are strong indications that it must have existed there as well.

* * *

However, parallel to all this, there was another truly remarkable phenomenon in ancient Arabia: vast numbers of nomads were literate and covered the desert rocks with their graffiti.[22] This is surprising since nomads do not usually have much use for writing, particularly in the days before there was a ready supply of cheap paper. Their societies are perfectly adapted to life without literacy, where memory is highly developed and communication is by word of mouth. In antiquity, writing was even less useful to nomads than it is today, since papyrus outside Egypt was relatively expensive; the desert did not provide palm-leaf stalks or sticks for incising; they had more urgent uses for the leather provided by their herds; and they used little or no pottery, since it was likely to get broken in the nomadic life, so sherds, which provided a common writing surface in settled areas, were also unavailable. The only support they had in abundance was provided by the rocks of the desert. However, for most people these are not much use for sending messages or recording information in a nomadic milieu.[23]

So why did the nomads of Arabia learn to read and write but apparently use these skills only for graffiti? At this distance of time it is impossible to be sure, but the following seems a likely hypothesis.[24] People in non-literate societies — i.e. those in which memory and oral communication serve the purposes for which we use writing — need to have very well-developed memories in order to store all the information which we would normally write down. This also helps them learn things relatively quickly and easily. In the desert, curiosity is a survival skill, for in a hostile environment a lack of curiosity can be fatal. I would suggest that if a nomad went to an oasis like Dedān, Taymāʾ, or Dūmah and saw a merchant writing a receipt or a letter, he might have asked "What are you doing" and, when told, might have said "Teach me to do that", simply out of curiosity.[25]

[22] For a more detailed exploration of this subject see Macdonald 2009*a* I: 74–96.

[23] See Macdonald 2009*a* I: 81, n. 102 on attempts to suggest that these graffiti had practical purposes.

[24] For a more detailed exposition of this hypothesis see Macdonald 2009*a* I: 78–82.

[25] For modern examples of this happening see Macdonald 2009*a* I: 78–79, 96–97.

FIGURE 6. *Graffiti carved in Greek by members of nomadic tribes in southern Syria (see Macdonald, Al Muʾzzin & Nehmé 1996: 480–485; Macdonald 2009*a *I: 76).*

One might have thought that while learning letter shapes was relatively easy, mastering the concept of distinguishing between consonants and vowels and using only the former was a more sophisticated and difficult process. And yet it does not seem to have been so. Indeed, we have examples of nomads in southern Syria who learnt to write their names in Greek, and therefore with vowels, as well as in their own alphabets where they used only consonants (Fig. 6).[26] In view of my earlier remarks on learning to read from inscriptions in South Arabia, it is interesting to note that of the nomads who carved the three Greek graffiti in Figure 6, the author of number 1 had learned his Greek letters from inscriptions (as seen in the shape of the *ēta*, like a capital "H"), while the authors of numbers 2 and 3 had learned from handwriting (as seen in their *ēta*s which lack the top right vertical stroke).

Having learnt to write, the nomad would return to the desert and no doubt show off his skills to his family and friends, tracing the letters in the dust or cutting them with a sharp stone on a rock. Because his nomadic society had no other materials to write on, the skill would have remained more of a curiosity than something of practical use, except for one thing. Nomadic life involves long periods of solitary idleness, guarding the herds while they pasture, keeping a lookout for game and enemies, etc. Anything that can help pass the time is welcome. Some people carved their tribal marks on the rocks; others carved drawings, often with great skill. Writing provided the perfect pastime and both men and women among the nomads seized it with great enthusiasm, covering

[26] For discussion of these texts see Macdonald, Al Muʾzzin & Nehmé 1996: 480–485; Macdonald 2009*a* I: 76–77.

the rocks of the Syro-Arabian deserts with scores of thousands of graffiti. The graffito was the perfect medium for such circumstances. It could be as short or as long as the authors wanted, and since they were carving purely for their own amusement they could say whatever they liked, in whatever order new thoughts occurred to them, and it did not matter if they made mistakes. When they tired of carving their own graffiti, they could wander off and vandalize someone else's, often by subtly altering the letters to make it say something different, or by adding something rude!

The introduction of writing to nomadic societies in Arabia probably happened many times and, in addition, individual nomads from one group no doubt passed on the skill to individuals from another group. We have evidence of this informal "teaching" process in a number of Safaitic graffiti, which simply list the letters of the alphabet. These are not in any traditional letter order, such as that used in the Phoenico-Aramaic alphabet (from which we get our ABC) or that used for the Ancient South Arabian alphabet (the *hlḥm*). Instead, they are ordered according to each person's perception of which letters had similar shapes (see Macdonald 2009*a* I: 86–87).

The result of this multiple introduction of writing to nomadic groups and its informal dissemination is that we find many different alphabets used by nomads to write graffiti. Because they had nothing but rocks to write on, writing did not penetrate their society and make it depend on literacy, and it remained simply a pastime, though, in these circumstances, this "pastime" was in fact a practical use for writing. We thus have the curious phenomenon of a non-literate society which retained its use of memory and oral communication for all important and practical matters, but in which the vast majority of the population must have been literate. The huge numbers of graffiti by equally huge numbers of individuals suggest that there must have been almost universal literacy among the nomads of the Syro-Arabian deserts over a considerable period. We are reminded of the example of the Tuareg mentioned above, among whom there is almost universal literacy in their own script (the Tifinagh) but who maintain an entirely oral culture and use their own alphabet purely for fun, employing foreign languages and scripts when they need writing for a practical purpose such as sending a letter.

A script that is used only for carving informally on rocks develops in a rather different way from those used for public inscriptions or for private documents on perishable materials, in a literate society. For a start, since the author is carving the text purely for his own amusement, he is not particularly concerned with whether or not it is comprehensible. The author knows what he means, and that is all there is to it. Thus, there is no incentive to develop a fixed direction of writing, separation of words, ways of showing vowels, etc. which are all things designed to help the reader. In these graffiti there are no spaces between words and no word-dividers, no vowels, and while in some of the scripts the text can run right to left or left to right (e.g. Thamudic B), in the others it runs vertically (e.g. Thamudic C and D), and in yet others it can go in any direction (Hismaic and Safaitic).

Using a script that is spread informally and employed purely to carve graffiti also has an interesting effect on the letter forms. In some cases, for instance, the same letter form can stand for completely different sounds in different scripts, in others a completely new form seems to have been invented, or adapted from another alphabet (see the script table in Macdonald 2009*a* III: 34). Because of the nature of the surfaces most of the letters can face in any direction and no letter is dependent on its stance for its identity.

The earliest firm date we have for the graffiti by nomads is the mid-sixth century BC, when a Thamudic B text mentions Nabonidus, king of Babylon.[27] Eight centuries later, the latest to be dated is a Thamudic D text (JSTham 1) giving the name and patronym of a woman buried at Ḥegrā (modern Madāʾin Sāliḥ) in AD 267, next to an epitaph in the Nabataean script (JSNab 17, see below). In between we have many Safaitic graffiti that mention the Nabataeans, the Romans, and other peoples. But while these scripts of the nomads continued to be used much later than those of the oases, they are thought to have died out by the fourth century AD, for reasons we cannot explain.[28]

* * *

However, there are other, more shadowy, dialects in pre-Islamic Arabia whose speakers rarely seem to have felt the need to write in them. I have suggested elsewhere that the language which I have called "North Arabian" (Macdonald 2009*a* III: 29–30) was made up of two

[27] This was discovered and photographed by Dr Muhammad al-Najem, director of the Taymāʾ Museum, some distance from the oasis. A photograph, but no reading, was published in al-Taymāʾī 2006: 90. It should not be confused with the Taymanitic inscriptions mentioning Nabonidus, referred to above.

[28] The reason for the assumption that they ceased to be used by, or within, the fourth century BC is simply the lack of any reference to Christianity in them. This is very unsatisfactory, but at present we have no other evidence.

mutually comprehensible dialect bundles, most strikingly distinguished by the form of the definite article: one, "Ancient North Arabian" (ANA) which used the definite article *h(n)-*, and the other, "Old Arabic",[29] which used *al-*. Needless to say, this is not the only feature distinguishing these groups, but simply the most convenient for the purposes of classification.

There is a very curious difference in the way these two groups of dialects were used. While, as we have seen, there are thousands of public inscriptions and graffiti in the ANA dialects, both in the settled oases and among the nomads, at present there appear to be only just under a dozen texts wholly or partially in Old Arabic before the sixth century AD, when we find the earliest inscriptions in the Arabic script (Macdonald 2008*a*). I say "appear to be" because we need to bear in mind that a large majority of the inscriptions in the ANA *scripts* are graffiti and when these consist solely of names it is, of course, impossible to identify the dialect or even the language spoken by the author.[30] Nevertheless, it has to be said that in the fairly large numbers of these graffiti which contain more than names, only a handful show signs that their authors may have spoken Old Arabic rather than ANA dialects (Macdonald 2008*a*: 468).

The Old Arabic inscriptions which have been identified are written in a number of different alphabets: Ancient South Arabian; Dadanitic (the ANA script used in Dedān); in one or more of those used by the nomads; and in the Nabataean script. This shows that Old Arabic co-existed with ANA and was not simply a later development from it, but it also appears to mean that before the sixth century AD, Old Arabic was so rarely written that it did not have its own script.

The only attempt known so far to write Old Arabic in the Ancient South Arabian script is from the city of Qaryat al-Fāw on the north-western edge of the Empty Quarter, on a major trade route from Yemen to eastern Arabia and the Gulf. At certain periods, Fāw seems to have been dominated by the Arab tribes of Kinda, Madhḥig, and Qaḥṭān, so it seems likely that, at least during these periods, one or more dialects of Old Arabic were spoken there. The excavators reportedly found large numbers of inscriptions,[31] and from those published so far, it seems clear that the written language of the oasis was Sabaic, though there are reportedly a few texts in other languages and scripts, perhaps by visitors. Because Kinda, Madhḥig, and Qaḥṭān are famous Arab tribes in the literature of the Islamic period we tend to assume that all their members must have spoken Old Arabic. While this is a perfectly reasonable assumption, we should perhaps remember that it is no more than that, and that some at least of those at Fāw may have spoken Sabaic as their first, or at least their second language.

However, if a speaker of Old Arabic at Qaryat al-Fāw wanted to commission an inscription, the commonest, and therefore the easiest and probably cheapest, practice would be for it to be expressed in the Sabaic language and script, since the scribes would have been used to writing in these. If, however, the customer insisted that the language of the text should be Old Arabic, the scribe would have had to find a way to express this in the Sabaic script, because he was used to writing Sabaic, and Old Arabic, as a normally unwritten language, had no dedicated script of its own. One such inscription from Fāw has been published: the epitaph of ʿIgl bn Hfʿm.[32] There are other inscriptions carved in the Sabaic script but almost certainly in a North Arabian dialect, though they do not provide sufficient information to allow us to classify them as either ANA or Old Arabic.[33] We find specifically Old Arabic "intrusions" in texts in written languages in north-west Arabia. As mentioned above, at Ḥegrā, where Nabataean Aramaic was the written language, we have an attempt in the third century AD to write an epitaph in Aramaic helped out with Arabic words and phrases (JSNab 17) and, at the nearby oasis of Dedān, an honorific inscription in Dadanitic (JSLih 71) which also shows Old Arabic intrusions. These texts provide positive evidence for the hypothesis that Old Arabic was a purely spoken language at these periods, and this bolsters the negative evidence from the scarcity of texts purely in the Old Arabic language and the fact that it did not have a dedicated script.

Yet speakers of Old Arabic must have had the same needs as speakers of Ancient North Arabian: the need to commemorate the dead with a gravestone or epitaph, to honour an important person, to record religious acts, or to proclaim decrees if they were in government. Similarly,

[29] I am using the term "Old Arabic" in the same sense as Old English, Old French, Old Aramaic, etc. to refer to the group of dialects which are considered to be the ancestors of the various forms of the mediaeval and modern languages, in this case the spoken and written Arabic of the Islamic period. See further Macdonald 2008*a*: 464; 2009*a* III: 30.

[30] See Macdonald 2004: 493–494 on the impossibility of divining the language spoken by a person from the etymology of his/her name.

[31] Unfortunately, very few of these have been published, see Ansary 1982: 142–147.

[32] For bibliography see Macdonald 2008*a*: 467.

[33] These are the inscriptions I have called "Pure Undifferentiated North Arabian", see Macdonald 2009*a* III: 54–55.

if they were nomads, they must have felt the need to help pass the time while guarding the pasturing herds. Throughout the western two-thirds of the Peninsula, speakers of ANA and Ancient South Arabian created and adapted numerous scions of the South Semitic script family, and used writing for all kinds of purposes; so why did those speaking Old Arabic apparently stand aloof from this? At a later date, we know from the pre-Islamic Arabic poetry, that the presence of inscribed rocks in the desert was sufficiently well known to be used in poetic imagery,[34] and yet speakers of Old Arabic seem to have had no desire to add to their number.

As I mentioned earlier, in eastern Arabia inscriptions of any sort are very rare and those which have been found are in foreign scripts and, when they consist of more than names, in foreign languages: Akkadian cuneiform (Potts 1990, i: 305–307; 2010; André-Salvini & Lombard 1997; Glassner 2008), Aramaic (Healey & Bin Seray 1999–2000), Greek (Gatier, Lombard & al-Sindi 2002; Gatier 2007 and references there), and Ancient South Arabian (Robin 1994: 82–85). There are just over forty inscriptions, mostly gravestones, from Thāj, Qatīf, and other places in eastern Arabia, which are written in the Ancient South Arabian script (Sima 2002; Robin 1994: 80–81). However, the content of these texts is unfortunately so limited that at present it is impossible to identify the language precisely, though it may be a North Arabian dialect.[35] This uncertainty means that, in fact, we cannot even be sure what language or languages were *spoken* in eastern Arabia in the pre-Islamic period.

Is this apparent contrast between the west and the east of the Peninsula significant? Does it perhaps suggest that eastern Arabia was the home of Old Arabic in the first millennium BC and the first three centuries AD, and that by the third century AD speakers of Old Arabic dialects were moving west and in some way "replacing" the speakers of ANA dialects? Quite apart from the fact that such a hypothesis is based on an absence of evidence and must therefore be treated with suspicion, it would not explain why those speakers of Old Arabic who were nomads did not take over the ANA scripts still in use in the desert and why these were abandoned.

In South Arabia, the *musnad* and the *zabūr* remained in use until the mid-sixth century after which there is an epigraphic silence until the first inscriptions in Arabic some decades after the hijra. Yet, although from the early centuries AD onwards large numbers of Arabs had been settling in South Arabia, we do not have a single example from pre-Islamic Yemen of an attempt to write Old Arabic in the South Arabian script. In the north, all the ANA scripts, even those used by nomads, seem to have died out by the fourth century, and the Nabataean form of the Aramaic script was left as the only vehicle for writing in North Arabia.

* * *

From the mid-first millennium BC onwards, there are frequent references to "Arabs" living throughout the Fertile Crescent, from eastern Egypt, along the eastern Mediterranean, and throughout Syria and Mesopotamia, as well as in the Peninsula. These "Arabs" are described as having various ways of life: merchants, settled farmers, kings with large numbers of chariots, rulers of cities, brigands, and nomads. *Pace* Jan Retsö (2003: 577–626), I find it difficult to think of any characteristic other than a mix of language and some cultural elements which would identify — both to themselves and to the outside world — such geographically and socially diverse groups as the same people (Macdonald, 2009*c:* 304–307).

Among these, the Nabataeans are perhaps the best known and have left us the most extensive remains. The Nabataeans used Aramaic as their written language. Indeed, one of the first things we hear about them while they were still nomads in southern Jordan is that they "wrote a letter in Syrian characters [i.e. in the Aramaic script]"(Diodorus Siculus 19.96.1).[36] They gradually developed their own dialect of written Aramaic and their own form of the script. The use of Aramaic as a written language was already well established in northern Arabia, by the time they settled at Ḥegrā just north of Dedān, sometime in the first century BC, and brought with them the practice of using their form of the Aramaic language and script for writing. As mentioned earlier, Nabonidus seems to have introduced Imperial Aramaic as a prestige written language and script to Taymāʾ, and possibly other parts of north-west Arabia, in the sixth century BC. In Taymāʾ it seems to have ousted the local script but not, it appears, in Dedān where there are only a handful of graffiti in Imperial Aramaic[37] among the hundreds of texts in the local script, and no pre-Nabataean formal Aramaic inscriptions, as yet. From the first century AD, Nabataean Aramaic appears to have spread in north-west Arabia as a prestige written language. Indeed, long after AD 106,

[34] e.g. Labīd *Muʿallaqah* l. 2: ... *ka-mā ḍamina ʾl-wuḥiyyan silāmuhā* "... as though the stones bear writings".

[35] A slightly more informative text with what appears to be an interesting mixture of North Arabian and Aramaic will be published shortly, in Macdonald, forthcoming, *c*.

[36] On this, see most recently, Macdonald 2009*a* I: 97.

[37] See, for example, JSNab 224, 390; Naṣīf 1988: pl. CXXIV (a).

when the Romans annexed the Nabataean kingdom and renamed it *Provincia Arabia*, Nabataean Aramaic remained the primary and possibly the only local prestige written language of the region. The local scripts of Taymāʾ, Dedān, and Dūmah appear to have disappeared by this time and the only survivors of the South Semitic script in the north of Arabia were those used by the nomads, which clearly would have had no prestige at all.

This is neatly symbolized by a temple in a remote area of the Ḥijāz, called al-Ruwwāfah. After the annexation of the kingdom and its conversion into the Roman Province of Arabia, the language of government changed to Greek, but Nabataean Aramaic continued to be used as the written language for social and cultural purposes by many of its inhabitants, particularly in the more remote areas of the Province, like north-west Arabia. The temple at al-Ruwwāfah was built between AD 167 and 169 at the instigation of two successive governors of the Roman Province of Arabia, probably by a unit of the Roman army levied from the nomadic tribe of Thamūd (Macdonald 2009*a* VIII: 9–14). On it was placed a dedication to the emperors Marcus Aurelius and Lucius Verus in Greek, for the Roman side, and Nabataean as the local script of prestige, rather than one of the ANA scripts used by the nomads for graffiti. In consequence, it is doubtful whether the men in whose name the temple was built, or many of the people who passed it, would have been able to read the inscription in either language, but this was not the object of the exercise.

A different relationship between Nabataean Aramaic, Old Arabic, and an ANA script is seen a century later in Ḥegrā, where a man called Kaʿbū carved or commissioned an epitaph for his mother in Nabataean Aramaic. However, whoever composed it had a very poor grasp of the Aramaic language and had to fill out the gaps in his knowledge with Arabic words and phrases. It may have been he who added beside it her name and patronym in Thamudic D, one of the scripts of the nomads, perhaps as a sentimental reminder of their origins. If so, this could be one of the rare examples of a speaker of Old Arabic using an ANA script, though of course this can be no more than speculation dependent on certain assumptions.[38]

A similar situation occurs in a burial cave at Dayr al-Kahf in northern Jordan on the very edge of the desert (Macdonald 2006: 293, 296–298). Here, six large sarcophagi have been carved out of the rock. Around three of the walls, just below the ceiling, there is a neat (and this time correct) inscription in the Aramaic script of the Ḥawrān, explaining that a certain Ḥulayfū and his brothers, the sons of Awsū, made this tomb. But on each sarcophagus is the name of the deceased in one of the scripts of the nomads, Safaitic. The general announcement is in Nabataean, but the personal identification is in the script they used at home.

* * *

Ten years ago, I pointed out that the quite natural assumption that the Nabataeans spoke Arabic, even though they wrote in Aramaic, was not based on much sound evidence.[39] Shortly after this, however, some very compelling evidence was published in the legal papyri from the late first and early second centuries AD found in a cave in Naḥal Ḥever, to the west of the Dead Sea. These papyri are in Greek, Hebrew, Jewish Aramaic, and Nabataean Aramaic, and one of the very interesting points made by the editors is that in the Jewish Aramaic documents strings of Aramaic legal terms are followed by their equivalents in Hebrew; whereas in the Nabataean documents, strings of Aramaic legal terms are followed by their equivalents in Arabic. In contrast to the vast majority of the handful of Arabic loanwords previously identified in Nabataean, which come as one might expect from inscriptions in Arabia, these documents come from a Jewish community in the heart of the Nabataean kingdom, in what is now central Jordan.

If the Nabataeans had an established legal terminology in Arabic, this surely suggests that they used Arabic in their legal proceedings, even though the results were recorded in Aramaic. That such a practice is entirely possible is suggested by a comparison with the even more complicated situation in thirteenth-century England. In his famous book *From memory to written record: England 1066–1307*, Michael Clanchy analyses the interplay of languages in the establishment of the *veredicta* (jurors' answers, or "true sayings") in response to questions posed by the justices.

> First of all, the jurors were presented with the justices' questions (the "articles of eyre" technically) in writing in either Latin or French. They replied orally, probably in English, although their answers

[38] These are that Kaʿbū composed JSNab 17 himself, rather than employing someone else to do so; and that JSTham 1 was carved or commissioned by Kaʿbū, rather than it being a contemporary or later addition.

[39] It has sometimes been suggested that I claimed that the Nabataeans did *not* speak Arabic, but this is not the case. I merely pointed out that the evidence that they *did* speak Arabic evinced up to that time (2000), which mainly consisted of the etymological language of personal names, was not sufficient to support such a claim.

> were written down as *veredicta* by an enrolling clerk in Latin. When the justices arrived in court, the chief clerk read out the enrolled presentiments or *veredicta* in French, mentally translating them from Latin as he went along. On behalf of the jurors, their foreman or spokesman then presented the same answers at the bar in English. Once the presentiments, in both their French and English oral versions, were accepted by the court, they were recorded in the justices' plea rolls in Latin. Thus, between the justices' written questions being presented initially to the jurors and the final record of the plea roll, the language in use had changed at least five times, although it begins and ends with writings in Latin. ... If the oral English statement, which [the foreman of the jurors] presented at the bar, deviated in any detail from the written statements [in Latin and French], the jurors faced imprisonment.[40]

Compared to this, the proposed situation in which Nabataean legal proceedings could have taken place in Arabic but with all records being made in Aramaic, would be relatively simple.

In the late fourth century AD, Epiphanius of Salamis famously reported that the Nabataeans in Petra and Elusa sang hymns in Arabic. I also suspect (though it is unprovable) that the two lines of rhetorical Arabic included in the Nabataean inscription at ᶜĒn ᶜAvdat, which records an offering to the deified king Obodas, are a quotation from the Nabataean liturgy in praise of him. Religious liturgies tend to be extremely conservative in language, as in much else, as exemplified by the use of Latin in the Roman Catholic church until the Second Vatican Council, with a resurgence under Benedict XVI, or the continued use of Byzantine Greek and Old Church Slavonic in the Eastern Orthodox churches. The use of an ancestral language in Nabataean religious practice would not therefore be at all unlikely. If the fact that their contemporaries referred to the Nabataeans as "Arabs" means that they were Arabic-speakers (Macdonald 2009*c*: 307–310), this ancestral language is likely to have been Arabic.

If this is correct — and it is a big "if" — one might then envisage a society, at least in the southern parts of the Nabataean kingdom and later of the Province, in which the language of communication in everyday life was Arabic, in which the religious liturgies, and possibly literary works, were in Arabic and passed down orally, in which face-to-face political, administrative, and legal activity was conducted in Arabic; but when records or written communication were needed they were made in Aramaic. As I have suggested above, a similar, though not exactly comparable, situation existed in mediaeval Europe up to the late thirteenth century, where government, administrative, and legal activities were conducted orally in one or more vernaculars, but the written records and official communications were generally in Latin.

Even as late as the mid-fourth century AD, it was still possible to find those who could compose near perfect Aramaic for inscriptions, as another epitaph (of AD 356) from Ḥegrā shows (Stiehl 1970). We also find Aramaic in the Nabataean Aramaic script still used in graffiti in north-west Arabia in the fourth and fifth centuries. But here we have to be careful. For these graffiti are extremely formulaic and the same handful of Aramaic words or phrases is almost always used in them: *šlm*, *dkr*, *br*, *b-ṭb*, etc. These could well be linguistic fossils used as ideograms[41] and do not tell us what language the author actually *spoke*, any more than our use of *requiescat in pace* on gravestones does. What is much more exciting, is that some of these graffiti also include *Arabic* words and phrases (see Nehmé, this volume; and al-Ghabbān & Nehmé, forthcoming).

* * *

In these inscriptions and graffiti we have disjointed snapshots of the continuous but uncoordinated development of the Nabataean Aramaic script into the Arabic script. For it is now clear that the Arabic script was not created *ex nihilo* or consciously adapted from

[40] Clanchy 1993: 207. The whole chapter from which this passage is taken is a fascinating exploration of the interplay of the different languages used in mediaeval England.

[41] As is *br* instead of *bn* in the pre-Islamic Arabic inscriptions of Namārah, Zebed, Jabal Usays, and Harrān. *Pace* Robin (2006: 331–332), the *r* is clear in each of these texts and it is not possible to read it as a final *n*. Robin says that he believes that these inscriptions "confondent probablement la graphie du *n* final avec celle du *r*, comme certaines inscriptions plus tardives (voir ill. 10, l'inscription du Wādī al-Šāmiya, et comparer *br* ligne 1, avec *bn*, lignes 3 et 4)" (ibid.). However, the inscription of Wādī al-Shāmiyyah is between one and three centuries later than these Old Arabic inscriptions and is thus rather a remote comparison. Moreover, the published photograph is not sufficiently clear to check the accuracy of the facsimile (p. 96, Fig. 7 here). But even if the latter is accurate, Robin has confused two quite different things. In the Wādī al-Shāmiyyah inscription, all the examples of *n* and *r* have very similar forms (with the exception of the *r* of *ʾarbaᶜīn* which is different from all the other examples of *r*) and this is clearly a feature of the author's "hand". He has not "confused" the shape of the two letters, any more than the authors of the Old Arabic texts did. Indeed, the latter all make a clear distinction between the shapes of *r* and final *n*.

another writing system,[42] but is simply the latest form of the Nabataean script (Healey 1990–1991; Macdonald 2009*b*). As the Aramaic language came to be used less and less in Arabia, and Arabic at last started to be used for writing, the Nabataean script came to be associated with Arabic rather than with Aramaic, which is why we think of its latest phase as the "Arabic" script.

Now, as I have explained above, developments in a script tend to be pushed by writing in ink, not carving on stone. What we see in the inscriptions and graffiti is therefore a mixture of these developments with memories of the calligraphic letter forms used in public inscriptions. This means that we have to assume an extensive, and possibly increasing, use of writing on soft materials in the Nabataean script throughout the fourth to seventh centuries, since only this could produce the transitional letter forms and ligatures we see first in the "Nabataean" or "transitional" graffiti of the fifth century (see Nehmé, this volume), then in the early Arabic inscriptions of the sixth and seventh centuries, and the earliest Arabic papyri of the mid-seventh. This would also explain how a more or less consistent and apparently widely understood system of dots to distinguish between letters with the same form, could already have been in use by the first dated Arabic papyrus of 22 AH and the first dated inscription of the Islamic period (24 AH).[43] While the letter forms and most ligatures would, presumably, have developed in the way they did, regardless of whether the script written in ink was being used to express the Aramaic or the Arabic languages or both, certain orthographic features, such as the use of *tāʾ marbūṭah* and perhaps the development of *lām-alif*, are likely only to have developed through the use of the script to write Arabic.

Yet — and this is a crucial point — despite the extensive use of writing with pen and ink implied by (a) the development of the Nabataean into the Arabic script; (b) the confident handwriting of the earliest Arabic papyri; and (c) the reports from the early Islamic period mentioning writing and documents, Arab culture at the dawn of Islam was fundamentally oral.

* * *

Gregor Schoeler has demonstrated this in a series of brilliant articles and books, and again in his paper in this volume. As he shows, writing in the early Islamic centuries was used for practical purposes, for letters or memoranda, for treaties, legal documents, etc. (Schoeler 2002: 21), but religious materials (with the eventual exception of the Qurʾān), poetry and literary prose, genealogy, and historical traditions were transmitted orally, with all that that entails for the gradual metamorphosis and "improvement" of the texts. This did not stop transmitters and scholars keeping often extensive notebooks as aides-memoire. But the "publication" of literary, historical, and religious matter was by oral transmission, not by the written word.

It is very doubtful that such a situation came about suddenly in the first Islamic century and therefore, although we have no direct evidence from the *Jāhiliyyah*, it seems safe to assume that this was a situation which early Islamic society in Arabia inherited. It is a situation that we, with our total dependence on literacy, may find difficult to comprehend, and it was one which was to change in subsequent centuries in Islamic society. The decisive event was surely the decision under the Caliph ʿUthmān (23/644–35/656) to fix the text of the Qurʾān by having it committed to writing. Naturally, there was opposition (Schoeler 2002: 33, 50, 54), and for many years after it was done, there were those who maintained that a fixed written form should be something unique to the Qurʾān and that to write down Traditions of the Prophet or interpretations of the Qurʾān was to put them on the same level as the Holy Book (Schoeler, this volume and references there). This could be interpreted as a rearguard action against Islam becoming a *written* culture, rather than continuing what we assume to have been a centuries-old tradition in which culture was published and transmitted orally, with writing reserved for non-cultural activities, such as administration and business.

The latter are the sorts of texts, which are usually written on perishable materials and so it is likely that the vast majority of these documents have indeed perished. Thus, one reason why Arabic appears to have been a purely spoken language, at least in late antiquity, may be because the written documents have not survived. However, their existence can be inferred from the way the Nabataean script changed and developed into what we think of as the "Arabic script", an evolution that could only take place as a result of extensive writing in ink over a very long period.

[42] As suggested by, for instance, Milik *apud* Starcky 1966: cols 932–934; Troupeau 1991; Briquel-Chatonnet 1997. See also the subtler and more nuanced approach of Robin (2006: 326–330), though this does not demonstrate *how* the Arabic script could have been adapted from the Syriac.

[43] The fact that the diacritical dots were used occasionally and not consistently at this period (see also Déroche, this volume), does not take away from the fact that the system by which the number and position of dots distinguished particular letters has, with a few exceptions, been widely accepted and understood from the earliest Islamic documents and inscriptions until the present day.

Gradually, the use of the Aramaic language must have declined, in some places more quickly than in others, and since the requirement for written documents presumably remained, they came to be written in the only language available in the communities concerned, Old Arabic. If legal and administrative activities had always been carried on orally in Arabic, the necessary Arabic technical terminology would already have existed, the only change being the realization that the text did not need to be translated into Aramaic before it could be written down.

What I am suggesting therefore is that, before and immediately after the rise of Islam, Arab culture was in all important respects fundamentally oral, as is that of the Tuareg today. We know from the early Islamic period that poetry was transmitted orally and that the transmitters were expected to polish and change it, i.e. that it remained a living, protean thing (Schoeler, this volume and references there). The other most important aspect of ancient Arab culture, genealogy, also depends on oral transmission. For traditionally, genealogy in the Middle East, as in many other regions, is not simply a historical record, but a way of defining personal rights and responsibilities and of explaining social and political circumstances.[44] For this to be possible, the details at certain points in the genealogy need to be flexible and to change in order to "explain" shifting relationships and political positions in the real world. In an oral society, where "fact" is the consensus of what a sufficient number of people think they remember, it is very important that the tribal genealogy be kept in "men's hearts", because once it is fixed in writing it becomes a historical document, and no longer a constantly developing way of explaining social and political relationships. I would suggest that it was exactly because the tribal genealogies had not been written down in the *Jāhiliyyah*, that those responsible for producing the ideological infrastructure of the early Islamic state were able to incorporate the Old Testament genealogies into those of the Arab tribes. This "proved" the long-held belief that the Arabs were also descendants of Abraham,[45] and thus had an ancient relationship with the one true God. At the same time, they were producing a unifying "ethnic" identity for the "Arabs", which had not existed before and which would distinguish them clearly from the conquered peoples, even when these became Muslims (see also Retsö 2006: 16). All this required that the Arab genealogies used had been transmitted orally, and so were fluid enough to be adapted. The finished construction, however, was eventually committed to writing and became to all intents and purposes a fixed definition of what it was to be an "Arab", i.e. someone who could trace his ancestry to a point in this genealogy.

* * *

There is one final irony. Alphabets of the South Semitic script family were used in Arabia for a millennium and a half. Some of them were extremely beautiful; all had the clarity of clearly distinguished letter forms and writing without ligatures. They had the right number of letters to represent all the consonants of the Arabic language. Yet, while these alphabets were flourishing, Arabic remained largely unwritten. Only when they had disappeared from North Arabia and were fading out in the South, did Arabic begin to be written, not in one of Arabia's own scripts but in a form of the imported Aramaic alphabet (Macdonald 2008*b*). This was far less suitable than the South Semitic scripts, for it had begun with only twenty-two signs and by now had only sixteen different letter forms to represent the twenty-eight consonants of Arabic.[46] Yet, while the South Semitic scripts had been confined to Arabia apart from an offshoot in Ethiopia, this revitalized form of the Aramaic alphabet — the Arabic script — became the vehicle of a vibrant literary culture and has been used for many different languages, from the shores of the Atlantic to the South China Sea.[47]

Sigla

CIS ii	*Corpus Inscriptionum Semiticarum*. Pars II *Inscriptiones Aramaicas continens*. Paris: Imprimerie nationale, 1889–1954.
JSLih	Dadanitic inscriptions in Jaussen & Savignac 1909–1922.
JSNab	Nabataean inscriptions in Jaussen & Savignac 1909–1922.
JSTham	Thamudic inscriptions in Jaussen & Savignac 1909–1922.

[44] See for instance Lancaster 1981: 24–35 for the Bedouin; and, for a quite different use of genealogy in Yemen, Dresch 1989: 176–179.

[45] See, for instance, the very interesting exploration of the origins and spread of this belief in Millar 1993; 2005: 301–313.

[46] These are (initial and medial, not final, forms): (1) ʾ (2) *b-t-ṯ-y-n* (3) *g-ḥ-ḫ* (4) *d-ḏ* (5) *r-z* (6) *s-š* (7) *ṣ-ḍ* (8) *ṭ-ẓ* (9) ʿ-*ġ* (10) *f* (11) *q* (12) *k* (13) *l* (14) *h* (15) *w* (16) *y*.

[47] The twelve languages for which the Arabic script has at one time been used are: Arabic, Farsi, Fulani, Hausa, Kurdish, Malay, Ottoman Turkish, Pashtu, Sindhi, Swahili, Urdu, and Uyghur. See Daniels 1997.

References

André-Salvini B. & Lombard P.

1997. La découverte épigraphique de 1995 à Qalʿat al-Bahrein: un jalon pour la chronologie de la phase Dilmoun Moyen dans le Golfe arabe. *Proceedings of the Seminar for Arabian Studies* 27: 165–170.

al-Ansary A.T.

1982. *Qaryat al-Fau. A Portrait of Pre-Islamic Civilisation in Saudi Arabia*. London: Croom Helm.

Beaulieu P-A.

1989. *The Reign of Nabonidus King of Babylon 553–534 B.C.* (Yale Near Eastern Researches, 10). New Haven, CT: Yale University Press.

Beyer K. & Livingstone A.

1987. Die neuesten aramäischen Inschriften aus Taima. *Zeitschrift der Deutschen Morgenländischen Gesellschaft* 137: 285–296.

1990. Eine Neue Reichsaramäische Inschrift aus Taima. *Zeitschrift der Deutschen Morgenländischen Gesellschaft* 140: 1–2.

Briquel-Chatonnet F.

1997. De l'araméen à l'arabe: quelques réflexions sur la genèse de l'écriture arabe. Pages 136–149 in F. Déroche & F. Richard (eds), *Scribes et manuscrits du Moyen-Orient*. Paris: Bibliothèque nationale de France.

Campbell Thompson R.

1931. *The prisms of Esarhaddon and Ashurbanipal found at Nineveh, 1927–8*. London: British Museum.

Caskel W.

1954. *Lihyan und Lihyanisch*. (Arbeitsgemeinschaft für Forschung des Landes Nordrhein-Westfalen, Geisteswissenschaften, 4). Cologne: Westdeutscher Verlag.

Clanchy M.T.

1993. *From memory to written record: England 1066–1307*. (Second edition). Oxford: Blackwell.

Cross F.M.

1986. A New Aramaic Stele from Taymāʾ. *Catholic Biblical Quarterly* 48: 387–394.

Daniels P.T.

1997. The Protean Arabic Abjad. Pages 369–384 in A. Afsaruddin & A.H.M. Zahniser (eds), *Humanism, Culture, and Language in the Near East. Studies in Honor of Georg Krotkoff*. Winona Lake, IA: Eisenbrauns.

Daum W., Müller W.W., Nebes N. & Raunig W. (eds)

2000. *Im Land der Königin von Saba. Kunstschätze aus dem antiken Jemen*. Eine Ausstellung der Staatlichen Museums für Völkerkunde München. Munich: Staatliches Museum für Völkerkunde München.

Degen R.

1974. Die aramäischen Inschriften aus Taimāʾ und Umgebung. Pages 79–98 in R. Degen, W.W. Müller & W. Röllig, *Neue Ephemeris für semitische Epigraphik*. ii. Wiesbaden: Harrassowitz.

Deputy Ministry of Antiquities and Museums

2007 *Mashrūʿ al-baʿthah al-athariyyah al-saʿūdiyyah al-almāniyyah al-mushtarakah li-l-tanqīb ʿan āthār taymāʾ*. Riyadh.

Déroche F.

(this volume). The codex Parisino-petropolitanus and the *ḥijāzī* scripts. Pages 113–120 in M.C.A. Macdonald (ed.), *The development of Arabic as a written language*. (Supplement to the Proceedings of the Seminar for Arabian Studies 40). Oxford: Archaeopress.

Diodorus Siculus/ed. and trans C.H. Oldfather & R.M. Geer.

1933-1967 *The library of history*. (Loeb edition). London: Heinemann/Cambridge, MA: Harvard University Press.

Dresch P.

1989. *Tribes, Government and History in Yemen*. Oxford: Clarendon.

Drewes A.J., Higham T.F.G., Macdonald M.C.A. & Ramsey C.B.
(forthcoming) Some absolute dates for the development of the Ancient South Arabian miniscule script.

Eichmann R.
2009. Archaeological evidence of the pre-Islamic period (4th–6th cent. AD) at Taymāʾ. Pages 59–66 in J. Schiettecatte & C.J. Robin (eds), *L'Arabie à la veille de l'Islam. Bilan clinique.* (Orient et Méditerranée, 3). Paris: De Boccard.

Eichmann R., Schaudig H. & Hausleiter A.
2006. Archaeology and epigraphy at Tayma (Saudi Arabia). *Arabian Archaeology and Epigraphy* 17: 163–176.

Ephʿal I.
1982. *The Ancient Arabs. Nomads on the Borders of the Fertile Crescent 9th–5th Centuries B.C.* Jerusalem: Magnes/Leiden: Brill.

Frantsouzoff S.A.
1999. Hadramitic documents written on palm-leaf stalks. *Proceedings of the Seminar for Arabian Studies* 29: 55–65.

Gadd C.J.
1958. The Harran Inscriptions of Nabonidus. *Anatolian Studies* 8: 35–92.

Galand-Pernet P.
1998. *Littératures berbères. Des voix. Des lettres.* Paris: Presses Universitaires de France.

Gatier P-L.
2007. Sôtélès l'Athénien. *Arabian Archaeology and Epigraphy* 18: 75–79.

Gatier P-L., Lombard P. & al-Sindi Kh.M.
2002. Greek inscriptions from Bahrain. *Arabian Archaeology and Epigraphy* 13: 223–233.

al-Ghabbān ʿA.I. & Nehmé L.
(forthcoming). [Inscriptions from the Darb al-Bakrah].

Glassner J-J.
2008. Textes cunéiformes. Pages 171–205 in Y. Calvet & M. Pic (eds), *Failaka Fouilles Françaises 1984–1988. Matériel céramique du temple-tour et épigraphie.* (Travaux de la Maison de l'Orient et de la Méditerranée, 48). Lyon: Maison de l'Orient et de la Méditerranée.

Hawkins J.D.
2000. *Corpus of Hieroglyphic Luwian Inscriptions.* i. *Inscriptions of the Iron Age.* Part 1: *Text: Introduction, Karatepe, Karkamish, Tell Ahmar, Maras, Malatya, Commagene.* (Studies in Indo-European Language and Culture, NS 8/1). Berlin: de Gruyter.

Hayajneh H.
2001*a*. Der babylonische König Nabonid und der RBSRS in einigen neu publizierten frühnordarabischen Inschriften aus Taymāʾ. *Acta Orientalia* 62: 22–64.
2001*b*. First evidence of Nabonidus in the Ancient North Arabian inscriptions from the region of Taymāʾ. *Proceedings of the Seminar for Arabian Studies* 31: 81–95.

Healey J.F.
1990–1991. Nabataean to Arabic: Calligraphy and script development among the pre-Islamic Arabs. Pages 41–52 in J. Bartlett, D. Wasserstein & D. James (eds), The Role of the Book in the Civilisations of the Near East. Proceedings of the Conference held at the Royal Irish Academy and the Chester Beatty Library, Dublin, 29 June–1 July 1988. *Manuscripts of the Middle East* 5: 41–52.

Healey J.F. & Bin Seray H.
1999–2000. Aramaic in the Gulf: Towards a Corpus. *Aram* 11–12: 1–14.

Jaussen A. & Savignac M.R.
1909–1922. *Mission archéologique en Arabie.* (5 volumes). Paris: Leroux/Geuthner.

Lancaster W.
1981. *The Rwala Bedouin Today.* (Changing Culures). Cambridge: Cambridge University Press.

Littmann E.
1904. *Semitic Inscriptions*. (Part IV of the Publications of an American Archaeological Expedition to Syria in 1899–1900). New York: Century.
1940. *Thamūd und Ṣafā*. Studien zur altnordarabischen Inschriftenkunde. (Abhandlungen für die Kunde des Morgenlandes, 25/1). Leipzig: Brockhaus.
1943. *Safaïtic Inscriptions*. (Syria. Publications of the Princeton University Archaeological Expeditions to Syria in 1904–1905 and 1909. Division IV. Section C). Leiden: Brill.

Macdonald M.C.A.
1989. Cursive Safaitic Inscriptions? A Preliminary Investigation. Pages 62–81 in M.M. Ibrahim (ed.), *Arabian Studies in Honour of Mahmoud Ghul: Symposium at Yarmouk University December 8–11, 1984*. (Yarmouk University Publications: Institute of Archaeology and Anthropology Series, 2). Wiesbaden: Harrassowitz.
2004. Ancient North Arabian. Pages 488–533 in R.D. Woodard (ed.), *The Cambridge Encyclopedia of the World's Ancient Languages*. Cambridge: Cambridge University Press.
2006 [2008]. Death between the desert and the sown. Cave-tombs and inscriptions near Dayr al-Kahf in Jordan. *Damaszener Mitteilungen* 15: 273–301.
2008*a*. Old Arabic (Epigraphic). Pages 464–477 in K. Versteeg (ed.), *Encyclopedia of Arabic Language and Linguistics*. iii. Leiden: Brill.
2008*b*. The Phoenix of Phoinikēia: Alphabetic reincarnation in Arabia. Pages 207–229 in J. Baines, J. Bennet & S. Houston (eds), *The Disappearance of Writing Systems: Perspectives on Literacy and Communication*. London: Equinox.
2009*a*. *Literacy and Identity in Pre-Islamic Arabia*. (Variorum Collected Studies, 906). Farnham: Ashgate.
2009*b*. ARNA Nab 17 and the transition from the Nabataean to the Arabic script. Pages 207–240 in W. Arnold, M. Jursa, W.W. Müller & S. Procházka (eds), *Philologisches und Historisches zwischen Anatolien und Sokotra. Analecta Semitica In Memoriam Alexander Sima*. Wiesbaden: Harrassowitz.
2009*c*. Arabs, Arabias, and Arabic before Late Antiquity. *Topoi* 16: 277–332.
(forthcoming, *a*). On the uses of writing in ancient Arabia and the role of palaeography in studying them.
(forthcoming, *b*). Towards a re-assessment of the Ancient North Arabian alphabets used in the oasis of al-ᶜUlā.
(forthcoming, *c*). What is the language of the Hasaitic inscriptions? The longest Hasaitic text so far discovered. *Atlal*.

Macdonald M.C.A., Al Muᵓazzin M. & Nehmé L.
1996. Les inscriptions safaïtiques de Syrie, cent quarante ans après leur découverte. *Comptes rendus de l'Académie des Inscriptions & Belles-Lettres*: 435–494.

Milik J.T.
1972. *Dédicaces faites par des dieux (Palmyre, Hatra, Tyr) et des thiases sémitiques à l'époque romaine*. Recherches d'épigraphie proche-orientale, 1. (Bibliothèque archéologique et historique, 92). Paris: Geuthner.
1978. Inscription Nabatéenne de Teima. Page 98, no. 70 in F. Baratte (ed.), *Un royaume aux confins du désert: Pétra et la Nabatène*. Catalogue de l'exposition au Museum de Lyon, 18 novembre 1978–28 février 1979. Lyon: Museum de Lyon.

Millar F.G.B.
1993. Hagar, Ishmael, Josephus and the Origins of Islam. *Journal of Jewish Studies* 44: 23–45.
2005. The Theodosian Empire (408–450) and the Arabs: Saracens or Ishmaelites? Pages 297–314 in E.S. Gruen (ed.), *Cultural Borrowings and Ethnic Appropriations in Antiquity*. (Oriens et Occidens. Studien zu antiken Kulturkontakten und ihrem Nachleben, 8). Stuttgart: Steiner.

Müller W.W. & al-Said S.F.
2001. Der babylonische König Nabonid in taymanischen Inschriften. Pages 105–122 in N. Nebes (ed.), Erstes Arbeitstreffen der Arbeitsgemeinschaft Semitistik in der Deutschen Morgenländischen Gesellschaft vom 11. bis 13. September 2000 an der Friedrich-Schiller-Universität Jena. *Neue Beiträge zur Semitistik*. (Jenaer Beiträge zum Vorderen Orient, 5). Wiesbaden: Harrassowitz.

Naʾaman N.

2008. The Suhu Governors' Inscriptions in the Context of Mesopotamian Royal Inscriptions. Pages 221–236 in M. Cogan & D. Kahn (eds), *Treasures on Camels' Humps. Historical and Literary Studies from the Ancient Near East Presented to Israel Ephᶜal*. Jerusalem: Hebrew University Magnes Press.

Al-Najem M. & Macdonald M.C.A.

2009. A new Nabataean inscription from Taymāʾ. *Arabian Archaeology and Epigraphy* 20: 208–217.

Nasif ᶜA.A.

1988. *Al-ᶜUlā. An Historical and Archaeological Survey With Special Reference to Its Irrigation System*. Riyadh: King Saud University Press.

Nebes N.

1992. New Inscriptions from the Barāʾn Temple (al-ᶜAmāʾid) in the Oasis of Mārib. Pages 160–164 in A. Harrak (ed.), *Contacts between Cultures*. Selected papers from the 33rd International Congress of Asian and North African Studies, Toronto, August 15–25, 1990. i: West Asia and North Africa. Lampeter: Mellen.

Nehmé L.

(this volume). A glimpse of the development of the Nabataean script into Arabic based on old and new epigraphic material. Pages 47–88 in M.C.A. Macdonald (ed.), *The development of Arabic as a written language*. (Supplement to the Proceedings of the Seminar for Arabian Studies 40). Oxford: Archaeopress.

Parpola S.

1987. *The Correspondence of Sargon II*. Part I: *Letters from Assyria and the West*. (State Archives of Assyria, 1). Helsinki: Helsinki University Press.

Pirenne J.

1956. *Paléographie des inscriptions sud-arabes: contribution à la chronologie et à l'histoire de l'Arabie du Sud antique*. i: *Des origines jusqu'à l'époque himyarite*. (Verhandelingen van de Koninklijke Vlaamse Academie voor Wetenschappen, Letteren en Schone Kunsten van België. Klasse der Letteren, 26). Brussels: Paleis der Academiën.

Potts D.T.

1990. *The Arabian Gulf in Antiquity*. (2 volumes). Oxford: Clarendon.

2010. Cylinder seals and their use in the Arabian Peninsula. *Arabian Archaeology and Epigraphy* 21: 20–40.

Retsö J.

2003. *The Arabs in Antiquity. Their history from the Assyrians to the Umayyads*. London: RoutledgeCurzon.

2006. The Concept of Ethnicity, Nationality and the Study of Ancient History. *Topoi* 14: 9–17.

Robin C.J.

1994. Documents de l'Arabie antique, III. *Raydān* 6: 69–90.

2006. La réforme de l'écriture arabe à l'époque du califat médinois. *Mélanges de l'Université Saint-Joseph* 59: 319–364.

Ryckmans J.

2001. Origin and evolution of South Arabian minuscule writing on wood. *Arabian Archaeology and Epigraphy* 12: 223–235.

al-Said S.F.

1999. Dirāsah taḥlīliyyah li-nuqūš liḥyāniyyah jadīdah. *Majallah Jāmiᶜat al-Malik Saᶜūd*, (al-ādāb 2), 11: 1–47.

(in press). Aramaic inscriptions from Taymāʾ, 2005 Season. In R. Eichmann, A. Hausleiter, A. al-Najem & S.F. al-Said, Taymāʾ — Autumn 2004 and Spring 2005. Second report on the Saudi Arabian-German Joint Archaeological Project. *Atlal* 20.

Sass B.

2008. Wadi el-Hol and the alphabet. Pages 193–203 in C. Roche (ed.), *D'Ougarit à Jérusalem. Recueil d'études épigraphiques et archéologiques offert à Pierre Bordreuil*. (Orient et Méditerranée, 2). Paris: De Boccard.

Schoeler G.

2002. *The genesis of literature in Islam. From the aural to the read.* Revised edition in collaboration with and translated by S.M. Toorawa. (The New Edinburgh Islamic Surveys). Edinburgh: Edinburgh University Press.

(this volume). The relationship of literacy and memory in the second/eighth century. Pages 121–130 in M.C.A. Macdonald (ed.), *The development of Arabic as a written language.* (Supplement to the Proceedings of the Seminar for Arabian Studies 40). Oxford: Archaeopress.

al-Shahri A.A.M.

1991. Recent Epigraphic Discoveries in Dhofar. *Proceedings of the Seminar for Arabian Studies* 21: 173–191.

1994. *Kayfa ibtadaynā wa-kayfa irtaqaynā bi-ʾl-ḥaḍārat al-insāniyyah min shibh al-jazīrah al-ʿarabiyyah. Ẓafār, kitābāt hā wa-nuqūsh hā ʾl-qadīmah.* [Privately published. Printed by al-Ghurair, Dubai].

Sima A.

2002. Die hasaitischen Inschriften. Pages 167–200 in N. Nebes (ed.), *Neue Beiträge zur Semitistik.* Erstes Arbeitstreffen der Arbeitsgemeinschaft Semitistik in der Deutschen Morgenländischen Gesellschaft vom 11. bis 13. September 2000 an der Friedrich-Schiller-Universität Jena. (Jenaer Beiträge zum Vorderen Orient, 5). Wiesbaden: Harrassowitz.

Starcky J.

1966. Pétra et la Nabatène. Cols 886–1017 in L. Pirot, A. Robert, H. Cazelles & A. Feuillet (eds), *Supplément au Dictionnaire de la Bible.* vii. Paris: Letouzey.

Stein P.

2005*a*. The Ancient South Arabian minuscule inscriptions on wood. A new genre of pre-Islamic epigraphy. *Jaarbericht van het Vooraziatisch-Egyptisch Genootschap "Ex Oriente Lux"* 39: 181–199.

2005*b*. Stein vs. Holz, *musnad* vs. *zabūr* — Schrift und Schriftlichkeit im vorislamischen Arabien. *Die Welt des Orients* 35: 118–157.

Stiehl R.

1970. A New Nabataean Inscription. Pages 87–90 in R. Stiehl & H.E. Stier (eds), *Beiträge zur alten Geschichte und denen Nachleben. Festschrift für Franz Altheim zum 6.10.1968.* ii. Berlin: de Gruyter.

al-Taymāʾī M.H.al-S.

2006. *Minṭaqat rujūm ṣaʿṣaʿ bi-taymāʾ. Dirāsah athariyyah maydāniyyah.* Riyadh: Wakālat al-wizārah li-l-āthār wa-ʾl-matāḥif.

Troupeau G.

1991. Réflexions sur l'origine syriaque de l'écriture arabe. Pages 1562–1570 in A.S. Kaye (ed.), *Semitic Studies in honor of Wolf Leslau. On the occasion of his eighty-fifth birthday November 14th, 1991.* ii. Wiesbaden: Harrassowitz.

Winnett F.V. & Reed W.L.

1970. *Ancient Records from North Arabia.* (Near and Middle East Series, 6). Toronto: University of Toronto Press.

Author's address

Michael C.A. Macdonald, Oriental Institute, Pusey Lane, Oxford, OX1 2LE.

e-mail michael.macdonald@orinst.ox.ac.uk

M.C.A. Macdonald (ed.), *The development of Arabic as a written language.* (Supplement to the Proceedings of the Seminar for Arabian Studies 40). Oxford: Archaeopress, 2010, pp. 29–46.

Mount Nebo, Jabal Ramm, and the status of Christian Palestinian Aramaic and Old Arabic in Late Roman Palestine and Arabia

Robert Hoyland

Summary
This paper discusses the status and distribution of Old Arabic and Christian Palestinian Aramaic in the provinces of Palestine and Arabia during the Late Roman period. It investigates the linguistic context of the sixth-century mosaic inscription in the church of St George at Mount Nebo, which has been said to contain a word in Arabic, and concludes that it is almost certainly in Christian Palestinian Aramaic. Another pre-Islamic text, which has been alleged to be in Arabic, a graffito from Jabal Ramm, is shown to be most likely in Nabataean Aramaic.

Keywords: Old Arabic, Christian-Palestinian Aramaic, Palestine, Arabia, Late Roman

There are three Old Arabic[1] inscriptions written in the Arabic script that are accepted as such by all scholars: one from Zebed, south-east of Aleppo, one from Jabal Usays, south-east of Damascus, and another from Ḥarrān, south of Damascus (Macdonald 2008: nos X–XII). Two further inscriptions, one from Umm al-Jimāl and one from Jabal Ramm, at the very northern and southern extremities of modern Jordan respectively, are on the possible list, but there is no agreed reading and they may well be Nabataean Aramaic rather than Arabic (see Macdonald 2008: nos XIII and IX respectively; and Appendix 2 below). There is one other contender, from Mount Nebo, also in Jordan, but unfortunately — and oddly, given the rarity of such texts — it has not received much attention. Scholars who have looked at it have either accepted or rejected its Arabic credentials without any formal discussion. Since other scholars, myself included, have simply accepted the opinion of either the "pro-Arabic" lobby or the "anti-Arabic" lobby without demur, it seems worth giving the text its due consideration and that will be the aim of this paper.

The object of our study is found in the church of St George in the modern village of Khirbat al-Mukhayyaṭ, ancient Nebo (now approximately 30 km west of modern Amman) where Moses is alleged to have looked upon the Holy Land. There are a number of fine mosaic floors in this church, which were completed in the 530s AD. One of them, in the sacristy on the south side of the central apse, contains just two words (Fig. 1). The word on the right (as one looks at the mosaic) gives a person's name, in Greek letters: *Saōla*. The word on the left is the one that concerns us here. The first editors, Saller and Bagatti (1949: 171), took it to be in the Arabic language and script, reading *bi-salām* "[rest] in peace". A decade later Milik (1959–1960: 159) penned a note on it, observing that though this reading was satisfactory as regards sense, it was difficult as regards palaeography, and he favoured taking it as Christian Palestinian Aramaic (CPA), reading: *nayyeḥ šawzeb* "[oh God] give repose [and] give salvation". A quarter of a century passed by without further comment and then two articles came along together, one in favour of Arabic and one in favour of CPA. The former, by Knauf (1984), merely restated that the word was in Arabic without providing any supporting argument, except to say that Milik's reading was not convincing. The latter, by Puech (1984*a*), offered a new suggestion, namely that the text was bilingual: the word on the left was simply the name Saōla written in CPA, but comprising many errors and innovations on the part of the mosaicists. Which solution should we prefer? There are two aspects to this question that require elaboration before an answer can be given: the linguistic context in which this artefact is situated and the palaeography of the word itself.

[1] By this I mean the Arabic language from the pre-Islamic period. Since we have so few texts it is difficult to be sure how similar/different it was to the Arabic of the early Islamic period or to what degree it constituted, in its written form, a fairly homogeneous entity. For what we do know about it see Müller 1982; Macdonald 2009*a* III: 48–57; 2008; Robin 2001.

Figure 1. *The Saōla mosaic at the church of St George, Khirbat al-Mukhayyaṭ, Mount Nebo. (Photograph by Dr Lihi Habas).*

a. 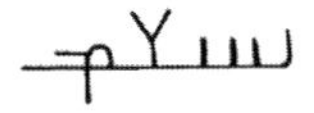

Actual letter forms of the word in the mosaic.

b. 

Puech's reading as it would be rendered in standard CPA script.

c. 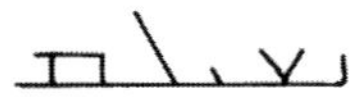

Bar-Asher's reading as it would be rendered in standard CPA script.

d. 

Milik's reading as it would be rendered in standard CPA script.

e. 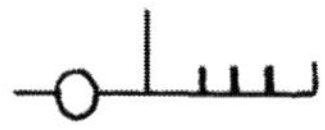

Compare how *bi-salām* / "in peace" would be written in 6th century Arabic script.

Figure 2. *Transcriptions of the suggested CPA readings of the left-hand word of the Saōla mosaic (Fig. 1).*

The linguistic context: Greek and Aramaic

It is important to point out that Nebo is only about 45 km to the east of Jerusalem, as the crow flies, and only a little to the south of the great Decapolis cities, such as Pella (modern Ṭabaqat Fiḥl), Gerasa (Jerash), Gadara (Umm Qays), Capitolias (Bayt Rās), Abila, and so on. In short, it was located within the orbit of major centres of Hellenic culture, where Greek was the key language. In late antiquity this pattern continued, with the Christian elite themselves becoming exponents of a Christian culture that was heavily Hellenized and conveyed principally in Greek. And it is not only at the high level of theological treatises, such as those penned by Sophronius, patriarch of Jerusalem (634–638), and Anastasius of Sinai (resident in Jerusalem in the 680s), and rhetorical displays, such as those of the Gazans Aeneas, Procopius, and Choricius (late fifth–early sixth century), that we see Greek in action. From papyrus finds we learn that in the sixth century the inhabitants of even such small provincial towns as Nessana and Petra conducted their everyday business in Greek.

Yet though Greek was certainly the official language in Late Roman Palestine and Arabia, many of those who lived there were Aramaic-speakers. To contemporaries Aramaic was simply "the language of the Syrians" (*ē tōn syrōn phōnē*), but there was an awareness that it came in many different dialects: "The inhabitants of the provinces of Osrhoene, Syria, Euphratesia, Palestine, and Phoenicia use the language (*phōnē*) of the Syrians, but all the same, the dialect (*dialexis*) of each exhibits many differences" (Brock 1994: 149, quoting Theodoret of Cyrrhus). In the region that interests us here the dialect is called Palestinian Syriac (*al-falasṭīniyyah*) by Bar Hebraeus (1890: 18), head of the west Syrian church in the thirteenth century, who says "it is the dialect of the people of Damascus, the mountains of Lebanon, and the rest of Inner Syria". Neither Theodoret nor Bar Hebraeus include Arabia in their list of Aramaic-speaking regions, as opposed to Palestine and Phoenicia, but this most likely reflects the province of Arabia's marginal status in their eyes, for modern excavations have uncovered a fair number of Aramaic texts from what was the Late Roman province of Arabia (modern northern Jordan and southern Syria). However, like Bar Hebraeus, modern scholars tend to use the epithet Palestinian to refer to the Aramaic dialect in use in the southern Levant (whether "Palestinian Syriac" or, more commonly now, "Christian Palestinian Aramaic"), presumably because Palestine was deemed a much more civilized and Christian province than Arabia.

Unlike the Syriac of northern Syria and Mesopotamia, CPA did not stand on equal terms with Greek, with its own original literature, but rather adopted a more subordinate role. Let us take, for example, the Greek inscription from Madaba that relates how, after a prolonged drought, the instruction of the bishop to construct a new cistern gave rise to a heavy rainfall. The inscription is principally in

Greek, but the response of the people of Madaba to the miraculous downpour is quoted in Aramaic: *goubba ba-goubba* "cistern by cistern", echoing the phrase *gebīm gebīm* that occurs in 2 Kings 3: 16 in connection with the rain miracles worked by Elijah, when God filled up all the cisterns/ditches in a particular valley upon Elijah's intercession. Thus, this text tells us not only that many of the people of Madaba in the sixth century AD spoke Aramaic, but also that they learnt their Bible in Aramaic as well; yet for formal writing purposes they had recourse to Greek. This need not mean, however, that they (or at least some of them) did not also *speak* Greek. We may have here a case of diglossia or code switching, where two languages are both spoken, but in different contexts. Typically, one will be deployed for more formal speech, the other being reserved for informal situations (prayers, blessings, jokes, taking notes, etc). As the editor of the inscription says, the composer of the text may have switched to Aramaic at this point "to record the spontaneous enthusiasm of the people of Madaba" upon seeing the miraculous downpour (Piccirillo 1981*a*: 311).

It is certainly true, however, that many of the people of this region were monolingual. Of some help in understanding the situation are the comments of the famous Spanish pilgrim Egeria, who travelled in the east around the 380s AD:

> In this province there are some people who know both Greek and Aramaic (*siriste*); but others know only one or the other language. The bishop may know Aramaic, but he never uses it. He always speaks in Greek and has a presbyter beside him who translates the Greek into Aramaic so that everyone can understand what he means (Wilkinson 1981: 146).

Egeria is describing a sermon given by the bishop of Jerusalem in the Church of the Anastasis, but we hear from Eusebius of Caesarea (d. 339) that the martyr Procopius (d. 303) had served in his home church in Scythopolis (another Decapolis city, some 80 km north of Nebo) as a translator into Aramaic (Garitte 1953: 255), and Epiphanius of Salamis (d. 403) speaks of those "who translate from one language to another at the readings as well as at the sermons" (1863: 3.2.21, col. 825). Furthermore, many an anecdote presupposes that a good proportion of the population of cities was made up of Aramaic-speakers, such as the tale of a woman who tells Porphyry of Gaza how she had never learnt Greek, but knew only Aramaic, and whose baby son starts speaking in Aramaic and is understood by all the onlookers (Grégoire & Kugener 1930: 52–55).

Nevertheless, Aramaic inscriptions in this region are few, and in its cities are almost never encountered. Madaba, for example, which has been thoroughly excavated for a second time, has not yielded a single Aramaic inscription, even though a number of new churches decked out with Greek texts were uncovered. Rather, Aramaic inscriptions crop up in small towns and villages, at a distance from major centres of settlement (see Appendix 1 and Fig. 4). This pattern suggests that only in the countryside could this language be used in public places and facilities, whereas in the urban centres, where the Hellenized elite held sway, Greek prevailed (despite the presence of a substantial number of Aramaic-speakers there too). Of course Greek could also be used outside cities — indeed, we have numerous Greek texts, especially graffiti and epitaphs, from very out-of-the-way places (e.g. Colt 1962: 137–193 [152 Greek inscriptions], 201–209 [ten Aramaic inscriptions] from Nessana and its environs) — but the point is that Greek was less *de rigueur* in such areas than in the cities and so it was possible in certain contexts for the Aramaic-speakers there to express themselves in their own tongue.

It was in particular for inscriptions of a more personal nature that CPA was employed. At the cemetery of Khirbat al-Samrāʾ, for example, 119 inscribed Christian tombstones survive, of which eighty-six are in CPA and only thirty-three in Greek (Humbert & Desreumaux 1998: 436–508 [CPA], 373–380 [Greek]). At Khirbat al-Mird, a number of CPA manuscripts were discovered, mostly containing Biblical texts and translations from Greek works, but among them was a letter (*c.* sixth–seventh century) written on papyrus by a monk named Gabriel to the head of the monastery there, asking for his prayers (Milik 1953: 533–539). Although at present unique, this letter may well have reflected a common occurrence, i.e. the use of CPA for simple correspondence. A second unique CPA document found nearby, in Wādī al-Nār to the east of Jerusalem, is a magical notebook, comprising various recipes and incantations, evidently for the private use of its owner (Baillet 1963). Even where the CPA text itself is public — inscriptions on mosaics and buildings — the function is very often personal: usually a blessing or prayer. For example, the CPA inscription that records the building of a hospital (*nosokomeion*) is by its nature a more formal text (no. 28 in Appendix 1 below and Fig. 3/a), but it, too, contains a blessing (on all those who enter).

For this reason we most often encounter CPA texts together with Greek and in a way that makes CPA's subordinate status and more personal function very clear (Levy-Rubin 1994). Thus at most of the churches where

3a. Inscription on limestone recording the founding of a hospital (Appendix 1, no. 28)

+ Glory to God! This hospital has been made in the days of our master Qayūmā + May whoever enters it profit by it (+ *tšbwḥtʾ l-ʾ[lhʾ] hdn nysqwmywn ʿbyd b-ywmwy d-mrn qywmʾ + kwl mn d-ʿll lh yhʾ myʿlʾ ʿl[wy]*).

3b. ʿAyn Suweinit, mosaic inscription (Appendix 1, no. 7)

May the Lord accept the offering of the priest Sīlā who made this cell, amen (*mrʾ ʾqbl qwrbnh d-qššʾ šylʾ d-ʿbd dʾ zʾwytʾ ʾmyn*).

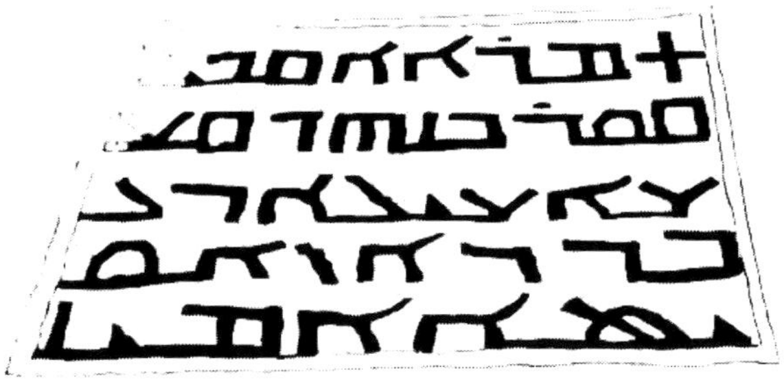

3c. Ḥayyān al-Mushrif, mosaic inscription (Appendix 1, no. 15)

O Lord, grant mercy to the priest Caius son of Eusebius who made this atrium (*mrʾ ʿybd rḥʾmy ʿl qʾšyšʾ qʾyʾ brh d-ʾswbwzw d-ʿbd drtʾ*).

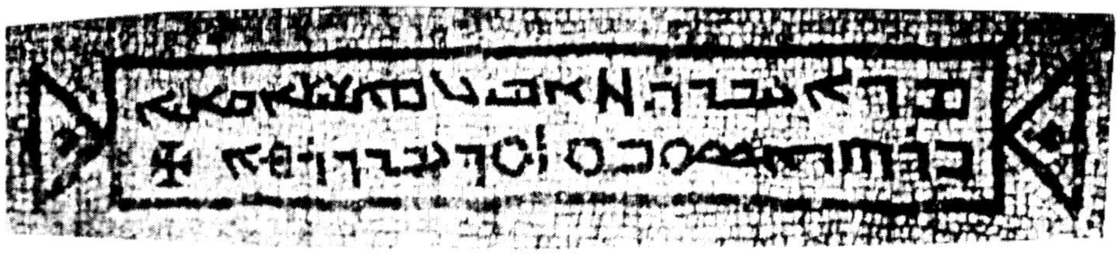

Figure 3a-c. *A selection of Christian Palestinian Aramaic inscriptions.*

3d. ʿAjlūn, mosaic inscription (Appendix 1, no. 21)

+ This holy place has been built by the hand of my master, the priest Sabinus, who has devoted his labour so that the Lord will assist him in the good work and so that the Lord will pardon him his sins and grant to him mercy, as also to its occupants and to all ages, amen (+ *hdn ʾtrʾ qdyšʾ ʾtqn ʿl ʾydwh d-mrn qšyšʾ sbyn mn d-lh ʾmn lʾwth d-mrʾ ʾsyʿnh b-ʿwbdh ṭbʾ d-mrʾ ʾšbwq lh syklth w-ʾʿbd ʿlwh rḥmn w-ʿl ʿmwrwh w-ʿl kwl ʿlmʾ ʾmn*).

3e. Quweismeh, mosaic inscription (Appendix 1, no. 19)

Lord Jesus Christ, bless this place and all of us who love Him/it, amen, for to Him recite (in this place) Stephen, ʿAbd Rayṭū and Ḥabība, amen (*mrʾ ysws mšyḥʾ ʾbryk hdn ʾtrʾ w-kwl mnn d-rḥm lh ʾmyn. d-lwth ʾmryn…*).

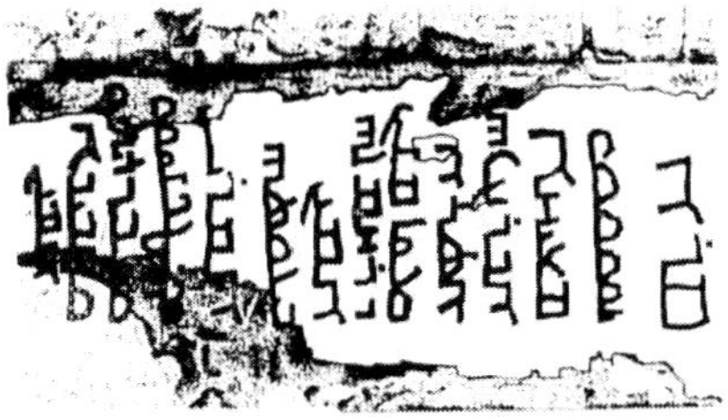

3f. ʿAyūn Mūsā, mosaic inscription (Appendix 1, no. 16)

May the Lord remember the good deeds of our master Gayyān the priest and the heirs who made… (*mrʾ yʿbd dkrwn d-ṭʾbn l-mrn gyʾn qšyšʾ w-l-yrwtwn d-ʿbdw mʾ…*).

3g. Khirbat Qastra, mosaic inscription (Appendix 1, no. 4)

Remember, o Lord, your servant Julian (*ʾdkr mrʾ ʿbdk lwlynh*).

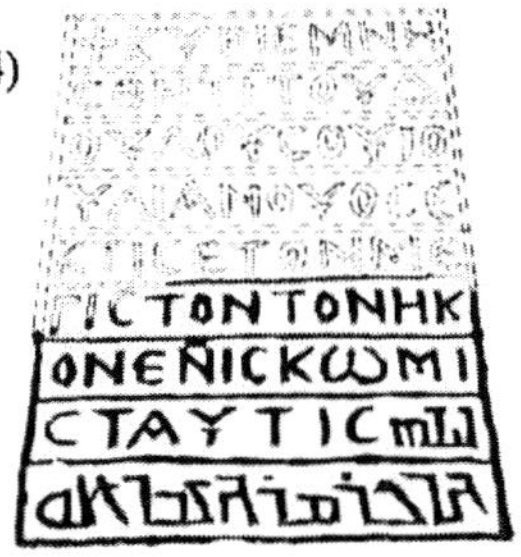

FIGURE 3d-g. *A selection of Christian Palestinian Aramaic inscriptions.*

CPA features in a mosaic, it does so alongside mosaics bearing Greek inscriptions that are situated much more prominently. The only exceptions to date are nos 7 (though the site has not been excavated), 15, and 21 in the list in Appendix 1 below (Fig. 3/b–d). The latter two are very interesting as their CPA inscriptions occur without accompanying Greek texts and are both placed in a very conspicuous location. CPA inscription no. 15 (at Ḥayyān al-Mushrif in modern north Jordan) is positioned directly in front of the southern entrance to the main hall in the centre of the monastery; CPA inscription no. 21 (at ᶜAjlūn in modern north-west Jordan, see Fig. 3/d) is framed by a very impressive geometrical design and lies west of the chancel screen, before the altar. Apart from these three cases, it is Greek that is used for the formal dedication whereas CPA was used for a private blessing or prayer. For instance, in the church of Quweismeh the Greek inscriptions are placed within the mosaic whereas the CPA one is outside it, though west of the chancel screen (no. 19 in Appendix 1 below and Fig. 3/e). The main Greek text, framed by a *tabula ansata*, proclaims the rebuilding of the church in AD 718 and records those who oversaw it; the CPA one asks Jesus to bless the site and remembers Stephen, ᶜAbd Raytū, and Ḥabībā. Similarly at ᶜAyūn Mūsā, there is a centrally-placed Greek inscription that conveys the formal dedication, while the CPA one is on the external border of the mosaic and transmits a personal request, calling on the reader to "remember the good deeds of our teacher Gayyān, the presbyter" (see no. 16 in Appendix 1 below and Fig. 3/f). The same pattern recurs at Khirbat al-Kursī, near Amman, and Evron near Haifa (see nos 18 and 1 respectively in Appendix 1 below). At Khirbat Qasṭra, CPA does feature in a bilingual inscription alongside Greek, but it gives only a heavily condensed version of the Greek wording (no. 4 in Appendix 1 below and Fig. 3/g). This pattern would seem to obtain also at the church of St George in Nebo. There, too, we find a number of prominently positioned Greek inscriptions that impart the important information: dates, names of donors and mosaicists, purpose, etc; and the non-Greek text is in a side room, again constituting a personal prayer, for the deceased Saōla (Piccirillo & Alliata 1998: 320, 327, 439–442).

Old Arabic

The status of Old Arabic in this region before Islam was even more marginal than that of Aramaic. For our knowledge of its existence we have nothing to go on but a handful of brief inscriptions (Macdonald 2009*a* III: 36–37, 48–54; 2008; Robin 2006), the Arabic words that feature in transliteration in the Greek papyri of Petra and Nessana (Daniel 1998; 2001; al-Ghul 2006), and a few cursory literary references (Retsö 2003: 591). These have been discussed a number of times and so there is no need for me to go over them again here. I would, however, like to make a couple of observations about the geographical distribution of these Old Arabic texts.

Firstly, with the exception of the text from Zebed, near Aleppo, which is only a short prayer to God — principally consisting of names — that has been added to a Greek-Aramaic bilingual building inscription (dated AD 512), all the Old Arabic texts and literary allusions to Arabic concern three specific areas (see Fig. 4). These comprise:

1. The Ḥawrān and basalt desert fringes: namely the Old Arabic inscriptions of al-Namārah, Jabal Usays, Ḥarrān and, possibly, Umm al-Jimāl (Macdonald 2008: nos VII, XI–XIII).

2. The Negev-Sinai region: namely the Arabic inscription of ᶜĒn ᶜAvdat/ᶜAyn ᶜAbadah (Macdonald 2008: no. VI) and the Nessana papyri.

3. The Petra–Madāʾin Ṣāliḥ region: namely Uranius (*c.* fourth–sixth centuries AD), who says that "Mōthō", usually identified with Muʾtah near Petra, means "death" in "the language of the Arabs" (*tē Arabōn phōnē*), and Epiphanius of Salamis, who notes that the people of Petra sang hymns to their virgin goddess "in the *Arabikē dialektos*" (for both references see Retsö 2003: 591; on the date of Uranius see Bowersock 2000); the Petra papyri; the Nabataean-Arabic transitional graffiti (see Nehmé, this volume).

Now these three areas are precisely where we find the main concentrations of Nabataean inscriptions (Hackl, Jenni & Schneider 2003: 719). Is this significant? Should we infer that under the influence of the Nabataeans, many of whom are generally held to have spoken an Arabic dialect (see Macdonald, this volume), Arabic became more widely spoken in these areas?

Secondly, there is almost no overlap between the distribution of CPA and pre-Islamic Arabic inscriptions. There are no CPA texts at all found in the areas of the Negev–Sinai or Petra–Madāʾin Ṣāliḥ, nor in the Ḥawrān (which, however, had its own Aramaic dialect and version of the Aramaic script; see Contini 1987: 31–36; Macdonald 2003: 44–46). Nabataean inscriptions have been discovered in the cities of Gerasa, Capitolias, and Madaba, but they are very few and most appear alongside Greek (Hackl, Jenni & Schneider 2003: 201–214, 719). Again, it is difficult to say whether, or in what way, this is significant. It is true that Nabataean inscriptions are chiefly located in the regions bordering on the desert and steppe whereas CPA texts are found nearer or within major agricultural areas. However, one would not want to imply that we are necessarily talking about two distinct populations or an exact coalescence of ecological and linguistic zones. Although at the level of the written word we just have the two scripts in the sixth century AD (besides Greek), namely Aramaic and Arabic, a traveller in these lands would have encountered, at the level of the spoken word, a variety of subtly differentiated dialects that often shaded one into another.

Thirdly, the area that has the most Old Arabic inscriptions, the Ḥawrān and the basalt desert fringes, is also the area where the best-documented pre-Islamic Arab dynasty, the Ghassanids, was based (Hoyland 2009). Should we connect these two facts? Two of these inscriptions are not only in the Arabic language, but also in what is recognisably the Arabic script. The Jabal Usays text (dated AD 528–529) is by a certain Ruqaym son of Muʿarrif,[2] sent to guard this important watering place south-east of Damascus on behalf of "king al-Ḥārith", the chief of the Ghassanids, who were a major ally of the Byzantine Empire. The Ḥarrān text (dated AD 568–569) is a bilingual Greek-Arabic inscription; it records the building of a martyrium for a certain Saint John by one Sharāḥīl son of Ẓālim, designated a phylarch in the Greek version of the text. The al-Namārah inscription also concerns an authority figure, a self-styled "king of all the Arabs" no less. Now it makes sense that it would be just such characters who would have promoted the use of Arabic, endowed as they were with a measure of power and resources and perhaps also with a sense of Arab identity (Hoyland 2007: 227–232). Since they had substantial dealings with Rome, it would also be natural to suppose that they had at least a rudimentary administration, and therefore scribes at their disposal. This is important, for as Michael Macdonald points out in this volume, changes in scripts tend to occur via the pen rather than the chisel. Thus, from the shift from Nabataean Aramaic script to Arabic script we must infer that there had been much writing of the Arabic language in the Nabataean script on parchments and papyri that very sadly have not survived. It is also noteworthy that the Ghassanids posed as champions of Christianity, as did the phylarch of the Ḥarrān inscription (since he commissioned the building of a martyrium), which might mean that certain Christian texts were translated into Arabic for the purpose of instructing new Arabophone converts (Hoyland 2008). Indeed, it was just such a purpose that stimulated the rise of the Coptic, Armenian, Georgian, and CPA scripts at about this time (namely the fourth–fifth centuries AD).

It is often thought surprising that Arabic was not written down for so long, but one must bear in mind that by and large most languages in the pre-modern world were not written down. The change from oral to written tends only to happen to a language when its speakers (or one particular group of them, which thereby make its dialect into the high form of the language) acquire sufficient common identity and sufficient political power to promote their own tongue. This began to take place in the case of Arabic when certain Arab tribes, through their interaction with the Late Roman Empire, gradually developed state structures and attained a measure of supra-tribal communal identity. This occurred principally in the Roman province of Arabia, which, in its original form of the annexed Nabataean kingdom, stretched from the southern outskirts of Damascus to the Ḥijāz (at least as far as Madāʾin Ṣāliḥ), and including the Negev desert, and it is therefore unsurprising that it is in this region that we find the earliest Arabic inscriptions. What muddies the picture is that pastoralists just outside this region would appear to have written for centuries in varieties of the South Semitic script (Macdonald 2009*a* III: 30–46; this volume). But in the pastoralist zones different rules apply. Most importantly they have time, space and, in this area in particular, large concentrations of the right sort of basalt and sandstone rock for carving. Moreover, the pressure to write in prestige languages, so strong in the settled lands (and which becomes overwhelming in cities) is not really felt out on the periphery.

Palaeography

If the word under discussion in the Nebo mosaic were in the Arabic script, it would be in a strikingly different form of it from the other sixth-century Arabic texts in the Arabic script. Assuming that the word is made up of the letters "b-s-l-m", the "b" (one vertical stroke) and "s" (three vertical strokes) would be standard for the Arabic script, but the putative "l" bifurcates as it rises[3] and the

[2] This reading of the name by Michael Macdonald (2009*b* and pp. 141–143 of this volume) is based on his new photograph of the inscription, see p. 141, fig. 1.

[3] Knauf would like to see it as an "l" and "ā" joined together (as in the printed form of the Arabic word). However, not only would this represent the only example of such a form in the whole history of Arabic writing (at least seeing it as an "l" with a bifurcation at the top does have parallels in later Arabic), but also medial "ā" was not generally written at this time. Knauf says there is one other example, namely the word *kātib* in the Umm al-Jimāl inscription, but see Appendix 2 below.

"m" has lines exiting from it downwards and leftwards. Oddly none of the proponents of an Arabic reading has made any comment about these startling letter forms. The only way to justify the unusual forms would be to argue that there already existed at this time an ornamented style of writing the Arabic script, an incipient Arabic calligraphy. Certainly the use of bifurcations on the apices of ascending letters (especially *l* and *aliph*) would become common in Arabic inscriptions by the ninth century AD (Grohmann 1957), but could such features have already existed, even if in a primitive form, in the sixth century? The "m" here does not resemble any later adorned form, but one could perhaps maintain that this was an experiment or a form that did not continue. And it is true that, irrespective of which script the word is written in, a degree of ornamentation is present, for the first vertical stroke and the baseline are executed in one sweeping line in red, whereas the rest of the letters were laid separately in black. The idea that there already existed the germs of an Arabic calligraphic style at this time is intriguing, but unfortunately we have no corroborating evidence and it is a lot to infer from just one word.[4]

Since, as we have shown above and as is clear from the inventory of CPA inscriptions given in Appendix 1 below, CPA was in common use in the area around Mount Nebo, and since an Arabic reading would be problematic, it seems more reasonable to take our text to be written in CPA. However, there is no very easy CPA reading either, and all the possibilities require some departure from the usual letter forms (see Fig. 2):

1. Puech (1984*a*) argues that the word should be read as an Aramaic rendering of the Greek *Saōla*. This would be satisfying in that the text would then be a simple bilingual one and it would conform to the evident symmetry of the mosaic as a whole (two animals either side of a central large tree, etc). However, it does demand that we postulate a high level of incompetence and/or innovation on the part of the mosaicists. He proposes that the first three vertical strokes are an open "s"; the next vertical stroke is either an open "w" that is linked to the "l" or a "y" that is substituting for a "w"; and at the end there is an inverted "ā" (Fig. 2/b).

2. Moshe Bar-Asher suggested to me the CPA reading "b-š-y-l-m" "in peace" (so the semantic equivalent of Arabic *bi-salām*), the *shwa* being written as a "y", which, he tells me, is known from CPA manuscripts. This is not impossible, but it does require accepting a number of unusual letter forms: the "š" as two parallel strokes rather than two strokes at right angles to each other, the "l" with a bifurcation at its apex, and the "m" with an unexplained downward stroke (Fig. 2/c).

3. The CPA reading of Milik is the only one to account for all the letter strokes, but it is by no means problem-free. The first four vertical strokes represent the letters "n-y-ḥ" of *nayyeḥ* "give repose" (Fig. 2/d).[5] Knauf argues that the second vertical stroke ought to be shorter if it is a CPA "y". In general, this is true, but the smooth sweep of the baseline indicates that the mosaicist had a concern for the aesthetic appearance of the text, which may have overridden other considerations. The initial vertical stroke is higher because it forms part of this baseline, which serves then to frame the whole text; the next three vertical strokes are exactly the same size, perhaps out of a desire for symmetry, which is a feature of the mosaic as a whole. One might also point out that the *ḥ* in CPA usually has a horizontal or diagonal crossbar (so that the letter looks like H or N), but it may well be that the baseline of the word made that unnecessary/impossible. The second word of Milik's reading, *šawzeb* "give salvation", begins with a "š", which Knauf feels is too tall for a CPA *š*, but there are other CPA inscriptions with a tall *š* (e.g. Fig. 3/c) and again aesthetic concerns may have played a part, since the letters are arranged either side of this central tall letter in the same way as the animals in the mosaic are arranged either side of the central tall tree. What had been taken by Saller-Bagatti and Knauf as an Arabic "m" is interpreted by Milik as the three letters "w-z-b", all joined together in one ligature. This does nicely explain all the different strokes (the circle is the "w", the downward stroke is the "z" and also the vertical side of the "b", which is completed by the upper leftward horizontal stroke and the baseline) and such agglomerations are attested in other CPA texts (e.g. Fig. 3/f: in *Gayyān*). Furthermore, the whole phrase equates effortlessly to the common Greek formula: *hyper sōtērias kai anapauseōs* "for the salvation and repose of".

It is true that to the Arabist or Arabophone it seems at first sight quite natural to read this word in the mosaic at St George's on Mount Nebo as *bi-salām* "in peace".

[4] Professor Irfan Shahid tells me that in his *Byzantium and the Arabs in the Sixth Century* ii/2 (Shahid 2010), of which I have not yet seen a published copy, he argues in favour of seeing the Mount Nebo mosaic inscription as an example of calligraphic Arabic, which was at that time being developed by the Ghassanids. It is, as I say, a tantalising theory, and it is only the lack of corroborating evidence and the prevalence of CPA in this region that holds me back from endorsing it.

[5] Sebastian Brock tells me that Milik's is definitely the most plausible solution, but he "would prefer to interpret the first word as *nyaḥ*, absolute from *nyāḥā*, i.e. Rest! (an exclamation rather than a wish)".

The vertical strokes of the "b" and "s", the vertical stroke of the "l" and the circular form of the "m" all appear to be present and to point effortlessly and ineluctably to an Arabic word. Yet, although in their broad movements these shapes recall Arabic letters, the details of the latter two, the bifurcation on the "l" and the extra strokes on the "m", would require us to postulate a calligraphic style of writing Arabic for which we have no evidence. It is not impossible — we have no evidence for any Arabic writing beyond the meagre scraps that I have mentioned above, even though we know there must have been dramatically more of it to propel the evolution of the Nabataean Aramaic script into what we call the Arabic script — and indeed it is an appealing and intriguing thought. However, as I have demonstrated above, Mount Nebo was in the sixth century situated in a region where the evidence for CPA is strong and the evidence for Arabic nil. Moreover, CPA is very frequently used alongside Greek, as in the St George inscription, whose author perhaps intended to illustrate that Saōla had been at home in both languages. These factors impel us to conclude that this word be seen as a CPA text and be struck off the list of pre-Islamic Arabic texts pending discoveries of calligraphic Arabic from the pre-Islamic period.

Appendix 1. An inventory of Christian Palestinian Aramaic inscriptions (see Fig. 4)

The following list includes the inscriptions listed in Bar-Asher (1977: nos 11–49) and Müller-Kessler (1991: 9–26), plus discoveries made since then. For the purposes of this list MI = mosaic inscription; RI = rock inscription; BI = building inscription; FI = funerary inscription. I select only those inscriptions which plausibly date to the fifth–eighth centuries AD and which, therefore, contribute to an understanding of the linguistic context of the Mount Nebo inscription. I only give the most recent/authoritative reference; for earlier literature the reader will have to refer to the works that I cite and Müller-Kessler 1991. Some of the inscriptions are very brief, occasionally only a name, and in this case it is not really possible to determine from the inscription alone whether it is in CPA or Syriac. However, Syriac is not generally used south of Emesa (except by travellers or émigré communities, e.g. Baramki & Stephan 1934), and therefore Aramaic texts by Christians of the southern Levant are most likely to be in CPA.

Phoenicia Maritima

1. Evron: MI (Jacques 1987).
2. Cabri: MI (Desreumaux 1987: 99–100).
3. Shlomi: MI (Dauphin & Kingsley 2003: 68 and n. 52).
4. Khirbat Qaṣṭra: MI (Puech 2001: no. 2).

Palaestina

5. Tell Masos: RI (Bar-Asher 1977: nos 144–145).
6. Umm al-Rūs: MI (Milik 1953: 527).
7. ʿAyn Suweinit: MI + RI: (Halloun & Rubin 1981; Puech 2001: no. 3).
8. Bethany: RI (Puech 2001: no. 4).
9. Khirbat al-Mird: RI (Milik 1961).[6]
10. Tell Yūnis: sherd (Bar-Asher 1975).
11. Khirbat Qaṣra: RI (Kloner 1990: 139/30*).
12. ʿAnab al-Kabīr: MI (Bar-Asher 2003).

Arabia

13. Gerasa: FI and two sherds (Milik 1953: 527–29, who says the tombstone was in the house of a Circassian in Jerash and is now in St Anne's Church in Jerusalem; although found in Jerash, it may not have originated there).
14. Rihab: MI (Piccirillo 1981*b*: 83).
15. Ḥayyān al-Mushrif: MI (al-Hasan 1996: 11–13).
16. ʿAyūn Mūsā: MI (Puech 1984*a*).
17. Khirbat al-Mukhayyaṭ/Nebo: MI (Milik 1959–1960: 159).
18. Khirbat al-Kursī: MI (Puech 1988).
19. Quweismeh: MI (Puech 1984*b*).
20. Khirbat al-Samrāʾ: FI x 86 (Humbert & Desreumaux 1998: 435–508).
21. ʿAjlun: MI (Puech 2003: no. 1).
22. al-Saʿnah: MI (Puech 2003: no. 2).
23. al-Burz: MI (Puech 2003: no. 3).
24. Qam: MI (Puech 2003: no. 4).
25. Umm al-Raṣāṣ: sherd (Puech 1989: 268–270).
26. Dayr ʿAdas: MI (Campanati 1995: 267).[7]
27. Dayr Mākir: BI (Naveh 1976: 102–103).

Unknown provenance

28. Inscription recording the building of a hospital (Milik 1953: 530–533, who says that it appeared in October 1952 on the antiquities market from an unidentified site in Transjordan).

---

[6] Numerous CPA papyri were also found here, including the papyrus letter mentioned above. The aforementioned magical papyrus notebook was found not far away as well.

[7] Campanati calls it Syriac, but Sebastian Brock has pointed out that it is CPA (2002: re 2.37). This text, along with nos 14–15, 21–24, and 27, are very important for showing the spread of CPA throughout Arabia, for it had been thought to be found only in the area of Amman and further south (e.g. see Millar 2009: n. 32).

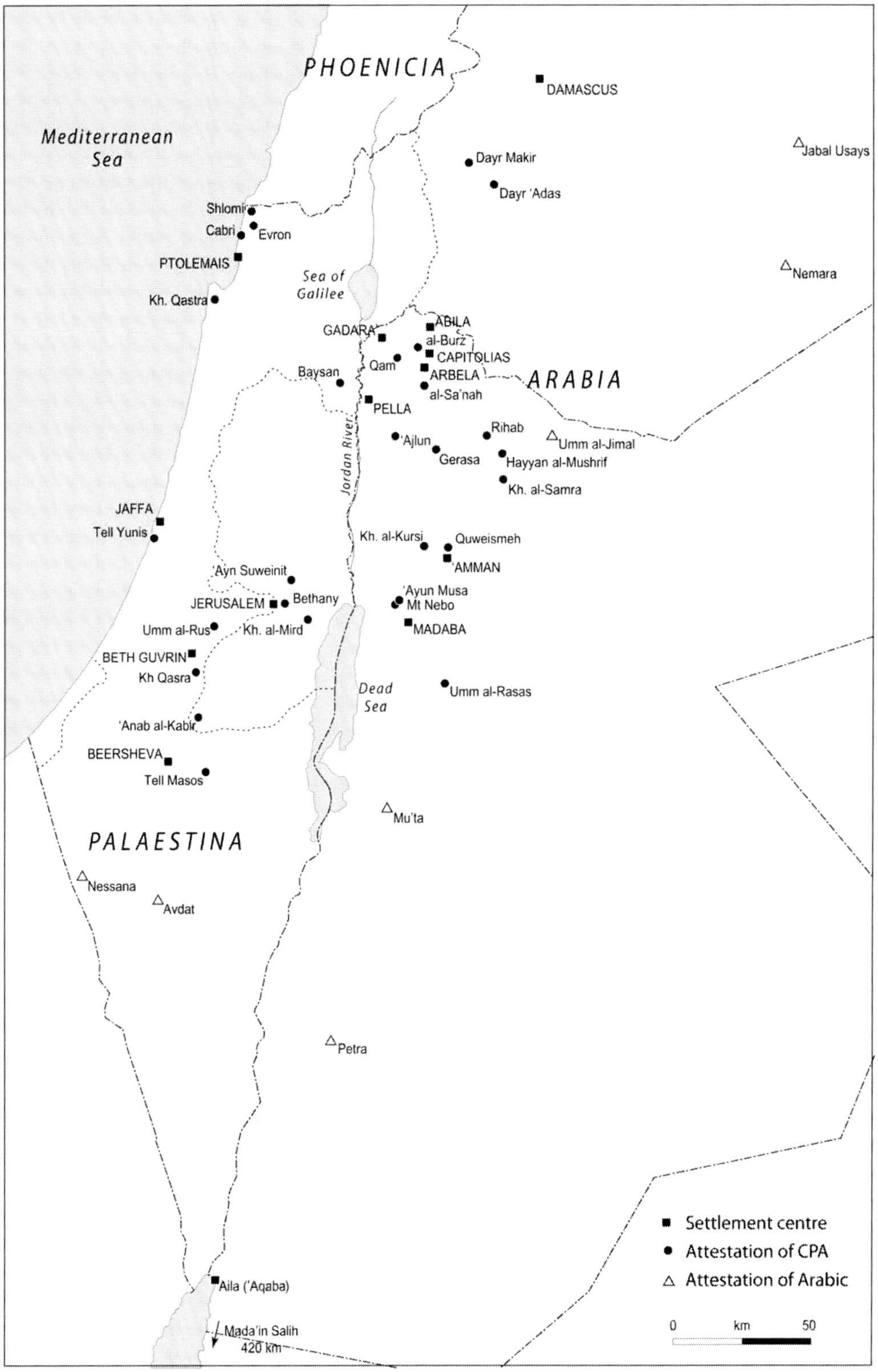

FIGURE 4. *A map of sites where evidence for CPA and Old Arabic exists.*

Figure 5. *The graffito from Jabal Ramm. (Photograph from Grohmann 1971: pl. I/1).*

29–30. Two amulets, one in Greek + CPA and one in CPA (Puech 1995: 299–302; 2001: no. 1).

31. Three inscribed lamps (Naveh 1988 — in the collection of R. Brown, Jerusalem; Naveh 1976: 103–104 — "acquired by the Department of Antiquities").

Appendix 2. A note on the Jabal Ramm and Umm al-Jimal inscriptions

Jabal Ramm (Fig. 5)

This text was etched into the plaster of a wall at the temple of Allāt in Ramm, ancient Iram (east of Aqaba, in modern south Jordan), and was found in the course of excavations at that site (Savignac & Horsfield 1935: 270). It has also been placed by many scholars on the pre-Islamic Arabic list; indeed, it has been proclaimed as "the oldest inscription in Arabic language *and* characters" (Gruendler 1993: 13, italics in the original). But now that we know much more about the different forms of Nabataean letters, we can see that it is written in the Nabataean Aramaic script (even if "evolved" or non-classical; see Nehmé, this volume). This also helps us with the reading, for Nabataean graffiti tend to deploy an extremely limited repertoire of expressions.[8] The text actually consists of two graffiti, which can be read as:

1. *...w br ᶜlyw ktb / ydh b-ʾrm*
   ...w son of ᶜAliyyū wrote (this) / with his own hand in Iram
2. *ḥbybw br ... b-šlm w-b-ṭ[b]*
   Ḥabībū son of ... in peace and wellbeing

A clear reading of the first graffito was hampered by the tendency of scholars to regard the curved groove at the end of the word on line 1 as a part of that word and to see its second letter as an "l" rather than an "evolved" *t*. This resulted in such readings as *klyṣ*/"plasterer" (Grimme 1936), *klnḍ* "a great number of coin" (Bellamy 1988), *ẖlyṣy* "of (the tribe of) Khulayṣ (al- ᶜUshsh 1973), and *kalbw* the proper name "Kalb" (Mascitelli 2006: 167). However, the curved groove should be seen as either part of the neighbouring Thamudic text or as an extraneous mark. And although the classical form of the Nabataean *t* is like the Greek letter Π, we do occasionally find it represented by a single inclining vertical stroke. It occurs thus, for example, in two texts from Saudi Arabia

[8] Grimme (1936) and Bellamy (1988: 370–372) did not take account of this and produced readings that, even if they fit the letter forms, do not conform to the phraseology of Nabataean graffiti.

(Sakāka, west of Ramm, and Umm Jadhayidh, south of Ramm), where it forms part of exactly the same phrase as is found here: "he wrote with his own hand" (Nehmé, this volume: figs 29, 45). This is a phrase that is found in Aramaic papyrus documents and the sense is: "he signed it". That the second line reads *yd-h b-ʾrm* has already been proposed by Macdonald (2009*a* III: 76, n. 171, addenda & corrigenda p. 13; 2008: 469, no. IX), followed by Mascitelli (2006: 169).

The second graffito also follows standard Nabataean phrasing: personal name plus greeting. I have given Ḥabību as the first name because it is known from other Nabataean texts, but the letter forms would also allow Ḥunaynū, as favoured by Grimme (1936). The rest of the name is uncertain. Grimme (1936) read "al-Muztalama/al-Maslamah"; al-ʿUshsh (1973) suggested "Umm al-Lammah". Neither is particularly compelling, but it is difficult to think of a convincing alternative. There would then seem to follow the usual Nabataean valediction, but inverted: "in peace and wellness" rather than "in wellness and peace". Previous readings took the last letter of this line to be a "k", but it has the curving S-like shape of the letter *ṭ*, and so "wellness" (*ṭb*) appears the most likely reading, especially after "in peace" (*b-šlm*).

These two authors evidently wanted to leave a record of their visit to this sacred place. They employed Nabataean letters and probably wrote, as far as they were concerned, in the Aramaic language. Although certain sectors of the population of the former Nabataean kingdom may well have spoken some Arabic dialect, as mentioned above and by Michael Macdonald (this volume), it is important to bear in mind that Aramaic was not only a formal language for inscriptions and documents, but was the spoken tongue of many, especially in centres of settlement. In the third century AD the Galilean rabbis Ḥiyyā the Great and Simeon ben Ḥalaftā even considered it worth their while making the journey to "Hegra of Arabia" (Madāʾin Ṣāliḥ) in order to "learn again" the meaning of some Aramaic words that they had forgotten.[9] That said, it is true that, judging from the words alone, the language of these graffiti could be either Aramaic or Arabic or a dialect in between; they are simply too brief to make a sure judgement. But in terms of their script and phraseology they fit well into the corpus examined by Laïla Nehmé (this volume).

Umm al-Jimāl (Fig. 6)

This is a very difficult text, as the stone was covered in plaster and was quite worn. A number of readings have been proposed (surveyed in Mascitelli 2006: 171–175), but none are very satisfactory. The task of making sense of this text has been exacerbated by the fact that a photograph was never published and everyone has relied on the transcription of Littmann (1929; 1949: 1–3), which is at least in part interpretative and has led to some fanciful readings (especially Bellamy 1988). It is in fact hard to say even what language this inscription is written in. For example, the Arabic word "ghafara" (forgive), which Littmann reads on the first line, only has a vertical stroke for its middle letter and so could at least as easily be read as "ʿabd", which is both an Arabic and an Aramaic word. Indeed, the final letter of this word, which Littmann takes to be a "r", is much more likely to be a "d", since in all the sixth-century Arabic inscriptions *d* and *r* are clearly distinguished. The *d* is always square (thus in the name *Saʿd* in the Zebed inscription and in the words *mafsad* and *baʿd* in the Ḥarrān inscription), as in the Umm al-Jimāl text (three clear examples), but the *r* is always curved. A lot of work remains to be done on this text before it can be read, and I hope to be able to do this in the near future, but in the meantime it seems worth publishing the late Geraldine King's photograph of it[10] so that other scholars might also try their hand at it, unencumbered by Littmann's own interpretation.

[9] Midrash Rabbah 79.7, re Genesis 33.19; Hebrew text in Theodor 1965: 946; English translation in Freedman 1951: 732–733 (who mistranslates Egra as Agora "market"). I am grateful to Dr Oded Irshai for this reference.

[10] I am grateful to her executors for making this photograph available to me.

Figure 6. *The allegedly Old Arabic inscription from Umm al-Jimāl. (Photograph by the late Geraldine King).*

References

Baillet M.
1963. Un livre magique en christo-palestinien à l'Université de Louvain. *Le Muséon* 76: 375–401.

Baramki D.C. & Stephan St.H.
1934. A Nestorian hermitage between Jericho and the Jordan. *Quarterly of the Department of the Antiquities of Palestine* 4: 81–86.

Bar-Asher M.
1975. A Palestinian Syriac inscription from Tell Yunis. *Haaretz Museum Annual* 17: 17–21. [In Hebrew].
1977. Palestinian Syriac Studies. PhD thesis, Deptartment of Linguistics, Hebrew University, Jerusalem. [In Hebrew; unpublished].
2003. The Syro-Palestinian Inscription from ᶜAnab al-Kabīr. *Tarbiz* 72: 615–620. [In Hebrew].

Bar Hebraeus/ed. A. Salihani
1890. *Mukhtaṣar al-duwal*. Beirut: Catholic Press.

Bellamy J.A.
1988. Two pre-Islamic Arabic Inscriptions revisited: Jabal Ramm and Umm al-Jimal. *Journal of the American Oriental Society* 108: 369–378.

Bowersock G.W.
2000. Two Greek Historians of pre-Islamic Arabia. Pages 123–134 in G.W. Bowersock, *Selected Papers on Late Antiquity*. Bari: Edipuglia.

Brock S.P.
1994. Greek and Syriac in Late Antique Syria. Pages 149–160 in A.K. Bowman & G. Woolf (eds), *Literacy and Power in the Ancient World*. Cambridge: Cambridge University Press.
2002. Some Basic Annotation to the Hidden Pearl: the Syrian Orthodox Church and its Aramaic heritage. http://syrcom.cua.edu/Hugoye/Vol5No1/HV5N1Brock.html.
Campanati R.F.
1995. Il mosaico pavimentale d'epoca umayyade della chiesa di S. Giorgio nel Deir al-Adas, Siria. Pages 257–269 in A. Iacobini & E. Zanini (eds), *Arte Profana e Arte Sacra a Bisanzio*. Rome: Argos.
Colt H.D.
1962. *Excavations at Nessana*. i. London: British School of Archaeology in Jerusalem.
Contini R.
1987. Il Hawran preislamico: ipotesi di storia linguistica. *Felix Ravenna* 133–134: 25–79.
Daniel R.W.
1998. Toponomastic Mal in P. Nessana and P. Petra Inv. 10 (Papyrus Petra Khaled and Suha Shoman). *Zeitschrift für Papyrologie und Epigraphik* 122: 195–196.
2001. P. Petra Inv. 10 and its Arabic. Pages 331–341 in I. Andorlini, G. Bastianini, M. Manfredi & G. Menci (eds), *Atti del XXII Congresso Internazionale di Papirologia (Firenze, 23–29 agosto 1998)*. i. Florence: Istituto Papirologico "G. Vitelli".
Dauphin C. & Kingsley S.A.
2003. Ceramic evidence for the rise and fall of a Late Antique ecclesiastical estate at Shelomi in Phoenicia Maritima. Pages 61–74 in G.C. Bottini, L. di Segni & L.D. Chrupcala (eds), *One Land Many Cultures; archaeological studies in honour of Stanislao Loffreda OFM*. (Collectio maior, 41). Jerusalem: Franciscan Press.
Desreumaux A.
1987. La naissance d'une nouvelle écriture araméenne à l'époque byzantine. *Semitica* 37: 95–107.
Epiphanius of Salamis/ed. J.P. Migne
1863. *Adversus Haereses*. (Patrologia Graeca, 42). Paris: Migne.
Freedman H.
1951. *Midrash Rabbah* (translation). London: Soncino.
Garitte G.
1953. Version géorgienne de Liber Annuus passion de S. Procope par Eusèbe. *Le Muséon* 66: 245–266.
al-Ghul O.
2006. Preliminary Notes on the Arabic Material in the Petra Papyri. *Topoi* 14: 139–169.
Grégoire H. & Kugener M.A.
1930. *Marc le diacre, vie de Porphyre évêque de Gaza*. Paris: Belles-Lettres.
Grimme H.
1936. A propos de quelques graffites du temple de Ramm. *Revue Biblique* 45: 90–95.
Grohmann A.
1957. The Origin and Development of Floriated Kufic. *Ars Orientalis* 2: 183–213.
Gruendler B.
1993. *The Development of the Arabic Scripts*. (Harvard Semitic Series, 43). Atlanta, GA: Scholars.
Hackl U., Jenni H. & Schneider C.
2003. *Quellen zur Geschichte der Nabatäer*. (Novum Testamentum et Orbis Antiquus, 51). Göttingen: Vandenhoeck & Ruprecht.
Halloun M. & Rubin R.
1981. Palestinian Syriac Inscription from "En Suweinit". *Liber Annuus* 31: 291–298.
al-Hasan ᶜA.Q.
1996. Aḍwāʾ jadīdah ᶜalā al-kitābah al-fusayfusāʾiyyah al-ārāmiyyah al-suryāniyyah al-masīḥiyyah al-muktashafah ḥadīthan fī mawqiᶜ Ḥayyān al-Mushrif – Mafraq". *Annual of the Department of Antiquites of Jordan* 40: 11–13. [Arabic section].

Hoyland R.G.
2007. Epigraphy and the Emergence of Arab Identity. Pages 219–242 in P. Sijpesteijn, L. Sundelin & S.T. Tovar (eds), *From Andalusia to Khurasan: Documents from the Medieval Islamic World*. Leiden: Brill.
2008. Epigraphy and the Linguistic Background to the Qurʾan. Pages 51–69 in G.S. Reynolds (ed.), *The Qurʾan in its Historical Context*. London: Routledge.
2009. Late Roman Provincia Arabia, Monophysite Monks and Arab tribes: a problem of centre and periphery. *Semitica et Classica* 2: 117–139.

Humbert J-B. & Desreumaux A.
1998. *Fouilles de Khirbet es-Samra en Jordanie*. Turnhout: Brepols.

Jacques A.
1987. A Palestinian-Syriac Inscription in the Mosaic Pavement at Evron. *Eretz Israel* 19: 54–56.

Kloner A.
1990. The cave chapel of Horvat Qaṣra. *Atiqot* (Hebrew Series) 10: 129–137. [English summary pp. 29*–30*].

Knauf A.
1984. Bemerkungen zur frühen Geschichte der arabischen Orthographie. *Orientalia* 53: 456–458.

Larcher P.
(this volume). In search of a standard. Dialect variation and New Arabic features in the oldest Arabic written documents. Pages 103–112 in M.C.A. Macdonald (ed.), *The development of Arabic as a written language.* (Supplement to Proceedings of the Seminar for Arabian Studies 40). Oxford: Archaeopress.

Levy-Rubin M.
1994. Society, Language and Culture in the Patriarchate of Jerusalem. Paper delivered at the fourth Late Antiquity and Early Islam conference on *Patterns of communal identity in the Late Antique and Early Islamic Near East.* [Unpublished].

Littmann E.
1929. Die vorislamische-arabische Inschrift aus Umm ij-Jimâl. *Zeitschrift für Semitistik* 7: 197–204.
1949. *Semitic Inscriptions: Arabic* (Publications of the Princeton University Archaeological Expeditions to Syria in 1904–5 and 1909, IV. D). Leiden: Brill.

Macdonald M.C.A.
2003. Languages, Scripts and the Use of Writing among the Nabataeans. Pages 37–56 in G. Markoe (ed.), *Petra Rediscovered: Lost city of the Nabataeans*. New York: Abrams.
2008. Old Arabic (Epigraphic). Pages 464–477 in K. Versteegh (ed.), *Encyclopedia of Arabic Language and Linguistics*. iii. Leiden: Brill.
2009*a*. *Literacy and Identity in pre-Islamic Arabia* (Variorum Collected Studies Series, 906). Farnham: Ashgate.
2009*b*. A note on new readings in line 1 of the Old Arabic graffito at Jabal Says. *Semitica et Classica* 2: 223.
(this volume). Ancient Arabia and the written word. Pages 5–28 in M.C.A. Macdonald (ed.), *The development of Arabic as a written language.* (Supplement to Proceedings of the Seminar for Arabian Studies 40). Oxford: Archaeopress.

Mascitelli D.
2006. *L'arabo in epoca preislamica: formazione di una lingua.* (Arabia Antica, 4). Rome: "L'Erma" di Bretschneider.

Milik J.T.
1953. Une inscription et une lettre en araméen christo-palestinien. *Revue Biblique* 60: 526–539.
1959–1960. Notes d'épigraphie et de topographie jordaniennes. *Liber Annuus* 10: 145–184.
1961. The Monastery of Kastellion. *Biblica* 42: 21–27.

Millar F.
2009. Christian Monasticism in Roman Arabia at the Birth of Muhammad. *Semitica et Classica* 2: 97–115.

Müller W.W.
1982. Das Altarabische und das klassische Arabisch. Pages 17–36 in W. Fischer (ed.), *Grundriss der arabischen Philologie*. i. Wiesbaden: Reichert.

Müller-Kessler C.
1991. *Grammatik des christlich-palästinisch-Aramäischen*. Hildesheim: Olms.
Naveh J.
1976. Syriac Miscellanea. *Atiqot* 11: 102–104.
1988. Lamp Inscriptions and Inverted Writing. *Israel Exploration Journal* 38: 36–43.
Nehmé L.
(this volume). A glimpse of the development of the Nabataean script into Arabic based on old and new epigraphic material. Pages 47–88 in M.C.A. Macdonald (ed.), *The development of Arabic as a written language.* (Supplement to Proceedings of the Seminar for Arabian Studies 40). Oxford: Archaeopress.
Piccirillo M.
1981*a*. La cattedrale di Madaba. *Liber Annuus* 31: 299–322.
1981*b*. *Chiese e mosaici della Giordania settentrionale*. Jerusalem: Franciscan Press.
Piccirillo M. & Alliata E.
1998. *Mount Nebo, new archaeological excavations 1967–1997*. Jerusalem: Franciscan Press.
Puech E.
1984*a*. L'inscription christo-palestinienne d'Ayoun Mousa (Mount Nebo). *Liber Annuus* 34: 319–328.
1984*b*. L'inscription christo-palestinienne du monastère d'el-Quweisme. *Liber Annuus* 34: 341–346.
1988. Les inscriptions christo-palestiniennes de Khirbet el-Kursi — Amman. *Liber Annuus* 38: 383–389.
1989. Une inscription sur jarre en christo-palestinien à Umm er-Rasas. *Liber Annuus* 39: 268–70.
1995. Deux amulettes palestiniennes, une en grec et une bilingue en grec-christo-palestinien. Pages 299–310 in H. Gasche & B. Hrouda (eds), *Collectanea Orientalia: arts de l'espace et industrie de la terre: études offertes en homage à Agnès Spycket.* (Civilisations du Proche-Orient. Série 1: Archéologie et environnement, 3). Neuchâtel: Recherches et Publications.
2001. Notes d'épigraphie christo-palestinienne cisjordanienne. *Revue Biblique* 108: 61–72.
2003. L'inscription christo-palestinienne du Ouadi Rajib-Ajloun et de nouvelles inscriptions christo-palestiniennes de Jordanie. Pages 317–325 in G.C. Bottini, L. di Segni & L.D. Chrupcala (eds), *One Land Many Cultures; archaeological studies in honour of Stanislao Loffreda OFM.* (Collectio maior, 41). Jerusalem: Franciscan Press.
Retsö J.
2003. *The Arabs in Antiquity. Their history from the Assyrians to the Umayyads.* London: RoutledgeCurzon.
Robin C.
2001. Les inscriptions de l'Arabie Antique et les études arabes. *Arabica* 48: 509–577.
2006. La réforme de l'écriture arabe à l'époque du califat médinois. *Mélanges de l'Université de St Joseph* 59: 319–364.
Saller S.J. & Bagatti B.
1949. *The Town of Nebo (Khirbet el-Mekhayyat): with a brief survey of other Christian monuments in Transjordan.* (Publications of the Studium Biblicum Franciscanum, 7). Jerusalem: Franciscan Press.
Savignac M-R. & Horsfield G.
1935. Le Temple de Ramm. *Revue Biblique* 44: 245–278.
Shahid I.
2010. *Byzantium and the Arabs in the Sixth Century.* ii/2. Washington, DC: Dumbarton Oaks Publications.
Theodor J.
1965. *Midrash Rabbah.* (Revised edition by C. Albeck). Jerusalem: Wahrmann.
al-ᶜUshsh M.
1973. Nashʾat al-khaṭṭ al-ᶜarabī wa-taṭawwuruh — al-khaṭṭ al-ᶜarabī qabl al-islām. *Annales Archéologiques Arabes Syriennes* 23: 54–84.
Wilkinson J.
1981. *Egeria's travels to the Holy Land.* Jerusalem: Ariel.

Author's address
Professor Robert Hoyland, The Oriental Institute, Oxford University, Oxford OX1 2LE, UK.
e-mail robert.hoyland@orinst.ox.ac.uk

.

M.C.A. Macdonald (ed.), *The development of Arabic as a written language*. (Supplement to the Proceedings of the Seminar for Arabian Studies 40). Oxford: Archaeopress, 2010, pp. 47–88.

A glimpse of the development of the Nabataean script into Arabic based on old and new epigraphic material

Laïla Nehmé

Summary

This contribution aims at presenting a corpus of epigraphic texts in a script that is "transitional" between Nabataean and Arabic. In order to establish this corpus, the author first collected all the texts which are dated to between the third and fifth centuries AD, whatever their origin. Secondly, "evolved" forms of characters, which occur in these dated texts, were sought for in undated ones. When identified, the texts in which these characters are found were included in the corpus. In Appendix 2, a sample of thirty-four texts is presented, together with photographs and facsimiles, of which fifteen are previously unpublished. The forms of the letters are analysed and those which can be termed "evolved" are identified and described.

Keywords: Arabia, Nabataean, Arabic, script

The last ten years have witnessed the discovery of increasing numbers of inscriptions of a type that was previously known only from a few texts recorded by earlier surveys in north-west Saudi Arabia.[1] These texts have usually been considered to be Nabataean. However, they show features which make them distinct from the Nabataean monumental and non-monumental inscriptions of the first century AD (called here, conventionally, "classical" Nabataean) as they appear mainly in Ḥegrā, Petra, and other sites in Jordan. Considering that these particular features appear to indicate a certain degree of development of the script, we have labelled these new texts, *faute de mieux,* "transitional", i.e. transitional between the Nabataean and Arabic scripts. This label is not entirely satisfactory because it can be understood only in the context of Nabataean and early Arabic epigraphy, but it is useful, at least provisionally, in order to identify, isolate, and describe these texts. It has been suggested that they should be called "Late Nabataean"[2] but this terminology would imply that they are closer to Nabataean than they are to the earliest examples of the Arabic script. This is probably true of some, but not all, of them.

The following contribution[3] has a very practical purpose and will not deal with the issue which is usually discussed when considering questions related to scripts in this field: the debate on the origin of the Arabic script. I consider, indeed, that at least in its early stages, the Arabic script did develop from Nabataean, not from Syriac, and that the corpus I shall be presenting is sufficiently convincing in itself to obviate presenting the historiography of the debate, the arguments for and against each theory, etc. Instead, I shall concentrate on what seems to me the most important element in the present state of our knowledge, i.e. the texts which have recently been found and the identification of the "transitional" ones, trying to answer the following question: which texts can be labelled as such and why?

A few major contributions have already dealt with the development of the Nabataean script into Arabic and I am very much indebted to them.[4] What follows is based on the work they have already done, which has shown me the way.

---

[1] For instance during the survey undertaken in 1962 by F.V. Winnett and W.L. Reed (on which see Winnett & Reed 1970).

[2] "Tardo-nabatéen," Robin 2008: 174.

[3] This contribution owes a lot to Michael Macdonald, who has not only read and corrected it as the editor of the *Supplement* would do, but has also made a lot of corrections, suggestions, and comments, including on the reading of some of the inscriptions. If I were to acknowledge each one of them, his name would appear under each paragraph. I would therefore like to express my warmest thanks to him for reviewing this text so carefully and making it ready for publication. I am however responsible for any mistakes which have remained in the text.

[4] Grohmann 1971; Gruendler 1993; Healey 1990–1991; Yardeni 2000, vol. B: 219–263. More recently, Macdonald 2009*a*; al-Muraykhī, in press. It should also be noted that this contribution deals only with epigraphic material written on *stone* and will therefore not consider the script on the papyri, which shows "evolved" forms of letters at an earlier date than in the inscriptions. In the analysis of the letters, I shall simply indicate when a similar form of a particular letter exists in the papyri.

The definition of "transitional" texts

The establishment of a new corpus of texts requires, at least at an initial stage, the use of objective criteria. Considering that I am interested in the development of the script from Nabataean to Arabic, I first decided to include in the initial corpus all the texts which can be dated to between the third and fifth centuries AD (I have labelled them "late Nabataean" texts), even if they do not necessarily show evolved features (see Appendix 1 for the list of texts dated to this period). The reason for choosing the fifth century is obvious: it is the last century before the appearance of the pre-Islamic texts from Syria in what is considered to be the early Arabic script, Zebed (AD 512), Jabal Says (AD 528), and Ḥarrān (AD 568).[5] The reason for choosing the third century is less obvious. One may ask, indeed, why not put the starting point of the corpus in the second century, which marks the end of the political independence of the Nabataean kingdom. However, this political event would be irrelevant to the use and development of the script. The main reason is the following: the third century is the period during which there are epigraphic texts which can be regarded as still being written in the "classical" Nabataean script as well as texts which show signs of a development towards something different.[6] This can best be illustrated by two texts, which are dated respectively to the beginning and the end of this century. The first one is *CIS* ii 963, from Wādī Mukattab in southern Sinai, dated to AD 206 (Fig. 23),[7] while the second one is UJadh 309, dated AD 295 (Fig. 48). One can easily see, on the facsimiles, that there are many more "evolved" characters in the text of AD 295 than there are in that of AD 206. Thus, in the text of AD 206, only final *y* and the *h* of *mʾh* have "evolved" forms, whereas in that of AD 295, this is true of — in order of their appearance in the text — the *y*, *d*, *š*, *r*, *ʾ*, *ḥ*, final *h*, final *t*, *m*, and *ʿ*.

All the letters that show "evolved" characteristics having been identified in the dated texts, the criteria used to include undated texts in the corpus of transitional texts were based on the fact that they contained such letter forms. Several hundred texts have been examined and a selection of those that have been included in the corpus, all of which are illustrated by photographs and facsimiles, is given in Appendix 2. The following methodology has been used: all the letter forms which appear in texts dated after AD 200 have been drawn and numbered individually, starting usually with the "classical" Nabataean form of the letter. The number of different letter forms does not exceed nine (for the *m*, for example) but there are usually no more than four or five. Each letter of each text which appears to contain evolved forms of letters or which is dated after AD 200 was then described in a database, using the numbers attributed to each of the various letter forms. It thus became possible to search for any form of any letter in initial/medial or final form (when relevant). Examining all the forms of all the letters led to the conclusion that only some of the letters had forms which could be considered as diagnostic and were therefore useful for the classification. The letters whose forms are not relevant *within this corpus* are the following:

— *b*: not only because the variations are small but also because these variations do not seem to be systematic;

— *z*: because the letter does not vary significantly;

— *ṭ*: there are three or four forms of the letter which can be reduced to two, defined basically by how the letter is traced: 1) like a Latin "S", i.e. a wavy line starting from the top, the letter remaining open on each side (as in UJadh 178, Fig. 40, or UJadh 219, Fig. 41); or open at the top and almost closed at the bottom (as in S 1, Fig. 29, and UJadh 375, Fig. 52); or 2) as a diagonal stroke terminated by a loop — either completely closed or open on the left — at the bottom of the letter (as in JSNab 18 line 2, Fig. 25). The first form is much more widespread than the second, of which there are only four examples in our corpus. Form 1, traced like an "S", is the one usually found in the cursive script of the papyri of the early second century AD, but there the bottom part of the letter forms a loop which is closed (Yardeni 2000: 247);

— *l*: because the looped form of the letter, which is the normal form in the "classical" Nabataean texts, where it coexists with the unlooped — one could say angular — form, does not occur in the texts dated after AD 200. The only attested form in the latter is the unlooped/angular form, which consists of the right and bottom sides of a rectangle and which does not vary significantly. This is also the form in the papyri (Yardeni 2000: 252);

— medial *n*: because the form of the letter is very consistent throughout the Nabataean and post-Nabataean period with, however, a general tendency to diminish the height of the vertical stem and give it the same height as

[5] On these texts, see Robin 2006: 330–338 with previous bibliography in notes.

[6] This does not mean that *all* the later texts are written in transitional characters.

[7] The figure numbers in the text do not start at 1 because priority has been given to the sequence of figures in Appendix 2, where the inscriptions are presented by their number in ascending order.

that of the *b*, with which it becomes easily confused (a good example is in QN 2, Fig. 28);

— *s*: because there are not enough examples of the letter in the texts. Where it does occur, however, the form of the letter is different from the "classical" Nabataean form. It is either open on the left, as in *CIS* ii 963 (Fig. 23), or closed but with a circular shape on the right and an angular shape on the left, as in UJadh 219 (= ThNUJ 84, Fig. 41). The form of the letter in the papyri is mostly open on the left but tends to become closed for ease of writing (Yardeni 2000: 256);

— ᶜ: because the letter has so many variations that it is almost impossible to trace a clear development of its form. Note, however, that the letter becomes more and more horizontal and tends to sit on the line, as in *ᶜnmw* in UJadh 219 (Fig. 41).

Having excluded these seven letters from our criteria, we are left with the other fifteen, which have forms which can clearly be termed "evolved" and which, in *combination* with each other, provide diagnostic criteria for the inclusion of a text in the corpus of transitional inscriptions.

— ᵓ (Fig. 1).[8] There are two forms of the letter, which can be considered as evolved forms, numbered 2 and 3 on Figure 1, while 1 is the most ordinary form of the letter in "classical" Nabataean. Form 3 can safely be considered as more evolved than 2. The best example of form 2 is ARNA.Nab 17 (Fig. 21), in *ᵓy*, *mᵓh*, and *khnᵓ*, while very good examples of form 3 appear in UJadh 309 (Fig. 48) in *ᵓwšw*, *ᵓlḥ{b/n}h*, *ktbᵓ*, and *mᵓt*. For *lām-alif*, see the end of this section.

— *g* (Fig. 1). The texts show two forms of the letter. The "classical" form is 1 (as in UJadh 360, Fig. 50), and it should be noted that this form is very similar to the evolved form of *ḥ* (form 3). However, there is another form of the letter (form 2), in which the diagonal stroke gets shorter and shorter and finally does not go down beyond the horizontal stroke (UJadh 3, Fig. 32) or only very little (Ar 19, Fig. 20). The last stage of the development of the letter, where the diagonal stroke stops at the horizontal, does not seem to be attested in the papyri.

— *d* (Fig. 1). The texts show three forms of the letter but only form 3 may be considered as "evolved" because it seems to appear only in texts from the third century onwards whereas 1 and 2 appear throughout Nabataean epigraphy, including in late texts. It should be noted, as

[8] Note that the letter shapes given in Figures 1,2, and 4 are a *synthesis* of all the letter forms, which were found in the texts and are not necessarily *exactly* how they appear in any particular text.

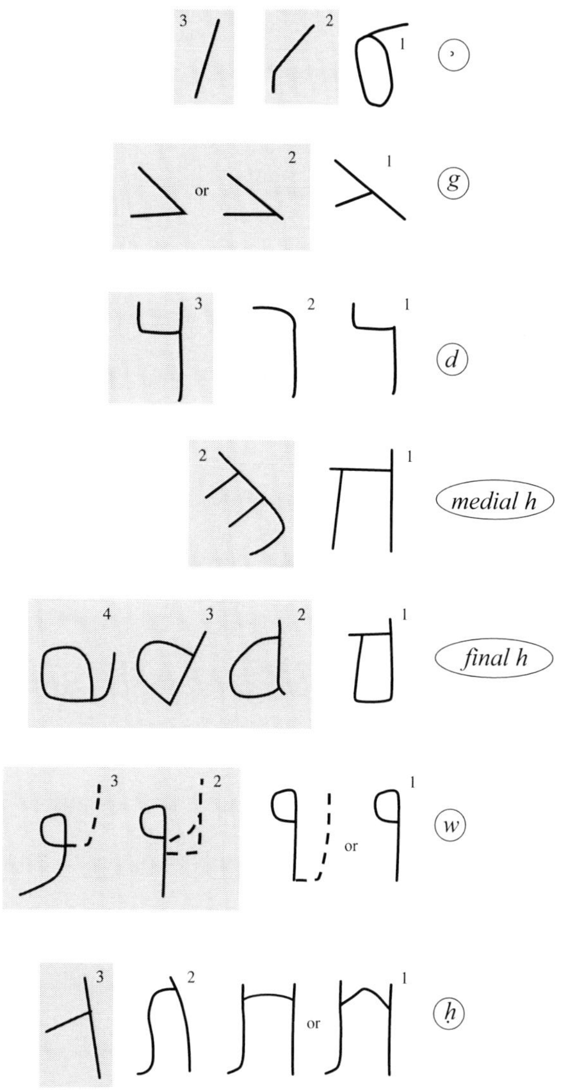

Figure 1. *The forms of* ᵓ, g, d, h, w, *and* ḥ.

M.C.A. Macdonald has pointed out, that form 3 is a far more "archaic" form, in terms of Aramaic epigraphy, than forms 1 and 2, which shows that an "evolved form" can develop the appearance of an "ancestral" form. It should also be emphasized that in all the transitional texts, *d* is clearly distinguished from *r*. Despite this, in a number of cases, a diacritical dot is placed above it (see the section "Dots on letters" below). A good example of *d* of form 3, with a dot, can be seen in UJadh 178 in the word *dkyr* (Fig. 40). In the table that contains the description of the characters for each text (Fig. 18), forms 1 and 2 are put together because they are sometimes very difficult to distinguish in the texts.

— medial *h* (Fig. 1). There are basically two forms of the letter but only form 2 appears in the late texts. Indeed, the "classical" form 1 disappears almost completely from the latter. The only examples are found in *CIS* ii 963 (Fig. 23), dated AD 206, and *RES* 528, which is unfortunately known only from a copy (Jaussen, Savignac & Vincent 1905: no. 2, copy p. 239). One incongruous form, in which the letter is closed at its bottom, can be found in ARNA.Nab 17 line 5 (Fig. 21), where the *h* in medial position has a normal medial form but, as noted by M.C.A. Macdonald, with the base line continued under it from the preceding *k*. The best examples of form 2 of medial *h* can be found in *phmw* in UJadh 375 (Fig. 52) and in *yhwdʾ* in ʿUlā 1 (Fig. 54).

— final *h* (Fig. 1). The "classical" form of the letter in Nabataean is 1 while form 3 is a variation of form 2, which is the form usually encountered in the late Nabataean and in the transitional texts.[9] Form 4 is a variation of form 3. Form 3 appears for instance in UJadh 297, dated to AD 305–306, mentioned above (Fig. 45), as well as in the Namārah inscription and in UJadh 309 in *ʾlḥ{b/n}h* (l. 2) and *dnh* (l. 3) (Fig. 48). UJadh 266 (Fig. 44) is a very good example of form 2. Finally, form 4 is found much more rarely, for instance in UJadh 299 (Fig. 47).

— *w* (Fig. 1). There is no need to give examples of form 1, which is the commonest form in "classical" Nabataean. Forms 2 and 3 are variations of the same, with a general tendency for the letter to be ligatured from the right halfway up the vertical stem of the letter. Note that the form where the ligature meets the stem at the back of the loop is the one that permits the transition to form 3. The two forms which appear under 2 (one where the ligature is attached more or less halfway up the vertical stem of the *w*, and one where it meets the stem at the back of the loop) should probably have been distinguished in the database. UJadh 219 (Fig. 41) has very good examples of form 2, the most widespread form of the letter in the transitional texts, while the patronym in UJadh 90 (Fig. 36) is the best example of form 3.

— *ḥ* (Fig. 1). Form 2 is only a variation of form 1, which is the "classical" Nabataean form. It is attested only in the inscription from Wādī Mughārah in Sinai (NDGS 2) and in UJadh 19 (= ThNUJ 34, Fig. 35). One interesting form, which may show how the letter developed, can be seen, as suggested by M.C.A. Macdonald, in the word *byrḥ* in JSNab 17, line 6 (Fig. 24). Form 3 is very widely attested in the transitional texts, as illustrated by UJadh 309 (Fig. 48), in *ʾlḥ{b/n}h* (l. 2) and *ḥd* (l. 5), and in UJadh 298 (Fig. 46), in the name *ʾlḥrt*.

— medial *y* (Fig. 2). The "classical" Nabataean form, 1, usually consists (from top to bottom) of an inclined stem followed by a loop — with a strong curve — open on the left. In late and transitional texts, the curve of the loop tends to diminish, to be less wide (form 2), and finally disappears, leaving only a more or less diagonal stem. It

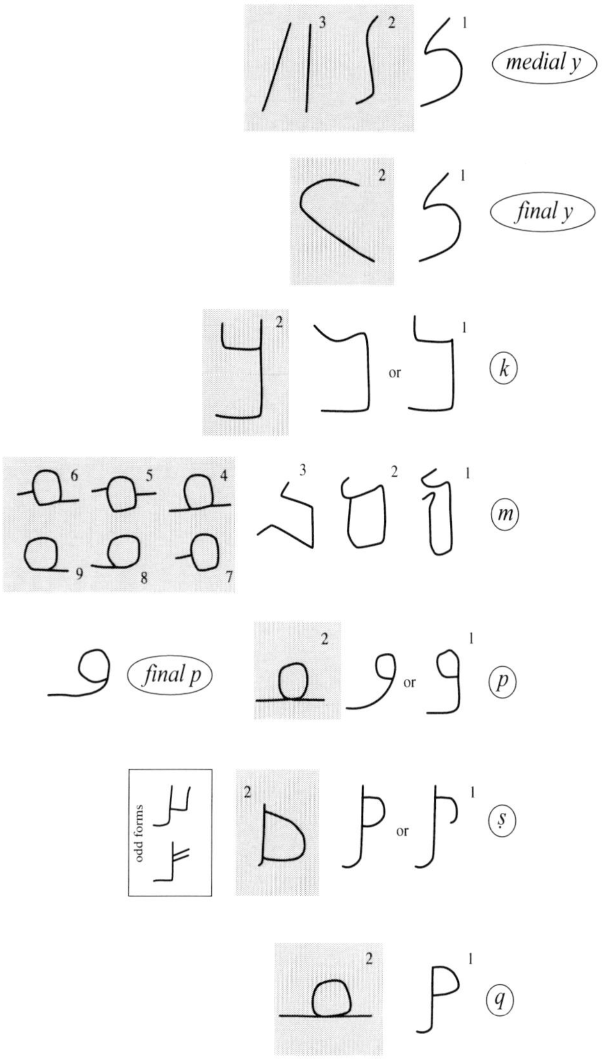

Figure 2. *The forms of medial and final* y, k, m, p, ṣ, *and* q.

[9] There are only four late or supposedly late texts which have form 1 whereas there are twenty-eight which have the other forms. In three of the former, several letters in the texts — not just the *h* — have "archaizing" forms. This is the case of JSNab 17, dated to AD 267 (Fig. 24), which has several other "archaizing" letter forms (*ʾ*, *š*, *t*, etc.), of the Stiehl inscription from Madāʾin Ṣāliḥ, dated to AD 356 (Fig. 31), and of the Fihrū text from Umm al-Jimāl, LPNab 41 (Fig. 26), dated to the third century. The fourth text is unpublished and comes from Madāʾin Ṣāliḥ. It may either be earlier than expected or an exception.

should also be noted that the letter starts to be joined to both the preceding and the following letters. Very good examples of form 2 can be seen in UJadh 309 (Fig. 48), in *dkyr* (l. 1), *ywm* (l. 4), and *tšᶜyn* (l. 6). As for form 3, which is probably a further step in the development of the letter, it can be best seen in UJadh 15 (= ThNUJ 30, Fig. 34), in *dkyr*, whereas in *šlym{n}*, in the same text, the *y* is between forms 2 and 3. Form 3 can also be seen in UJadh 178 (Fig. 40).

— final *y* (Fig. 2). There are basically two forms of final *y*. The first one is the same as medial form 1 but the transitional texts contain almost exclusively the final *y* which is shown as form 2. Good examples can be seen again in UJadh 309 (Fig. 48), in *bly* and *šly* (l. 1), in *btšry* (l. 5), as well as in UJadh 405 (Fig. 53), in both names. See also UJadh 3 (Fig. 32).

— *k* (Fig. 2). Form 1 and its variants are much less widespread than form 2,[10] which is very similar to the developed form of the *d*, except for the horizontal line at the bottom of the letter, which does not occur in the *d* but is an essential part of the letter in the *k*. M.C.A. Macdonald notes that as with *d*, the top of form 2 is typical of Imperial Aramaic *k*, although there the "tail" is a straight diagonal line. Examples of form 2 of *k* can be found in UJadh 15 (Fig. 34), UJadh 90 (Fig. 36), etc. There are not enough examples of final *k* to make any comment.

— *m* (Fig. 2). This letter is complicated to analyse because of the variations in the developed form of the letter. The "classical" Nabataean form is represented by forms 1–3 while the developed form is represented by forms 4–9.

Only sixteen texts in our group contain a letter *m* of forms 1, 2, or 3. In some of these texts, both the "classical" *and* developed forms of the letter occur in the same text. This is the case, for instance, in UJadh 10 (= ThNUJ 38, Fig. 33), where the *m* in *šlymw* and *yᶜmrw* is of form 7 and is ligatured to the left, whereas the *m* in *šmnw* is of form 1 and is ligatured from the right. It seems that both forms were perfectly familiar to the writer. The other texts — those in which there are only "classical" forms of this letter — are the following: one is a graffito from Madāʾin Ṣāliḥ, JSNab 18, (Fig. 25); four are formal texts, al-Namārah, JSNab 17, (Fig. 24), the Stiehl text, (Fig. 31), LPNab 41, (Fig. 26); four others are graffiti from Sinai[11] and are dated, apart from NDGS 2, to the first half of the third century; the same is true of B 3, from Boṣra (Fig. 22), which is dated to AD 230–231. Finally, there are two texts with a possible transitional character, one from al-ᶜUdhayb, north of al-ᶜUlā, and one from Umm Jadhāyidh, which contain only "classical" forms of *m*. They are not dated but have been considered as "transitional" because of the other letter forms in them (medial *h* of form 2 for al-ᶜUdhayb and possibly ᶜ for Umm Jadhāyidh). However, the Umm Jadhāyidh text at least (UJadh 360 = ThNUJ 62, Fig. 50), may have to be considered closer to "classical" Nabataean than we thought at first glance.[12] One last text, UJadh 172 (Fig. 39), is particularly interesting. If it is dated to AD 311–312 (see Appendix 2 for the reading of the date) and considering that it does not contain any transitional characters, it would show that the "classical" Nabataean script was still used in north-west Arabia in the fourth century AD.

Fifty-six texts — thus many more than in the above category — contain forms of *m* in which the body of the letter is close to a circle (nos 4 to 9 in Fig. 2). The letter is sometimes ligatured on both sides (forms 4–6) and sometimes on one side only (7–9).

Ligatures on both sides: when the letter is ligatured on both sides, the stem which makes the ligature is either at the bottom of the letter on each side (form 4), or in the middle of the letter on each side (form 5), or at the bottom of the letter on the right side and in the middle of the letter on the left side (form 6). Having examined all the examples recorded as form 4 (nineteen examples) and form 5 (sixteen examples) in the database, it appears that the position of the ligature depends partly on the form of the letter *m*: the flatter it is at its base, the more the ligature starts from the bottom of the letter (as in Ar 19, in the *m* in *ᶜmrw*, Fig. 20, or in UJadh 266, in the *m* of *ᶜmyyw*, Fig. 44); the closer it is to a circle, the more the ligature starts from the middle of the letter (as in UJadh 90, in *ynmw*, Fig. 36, or in UJadh 375, in *phmw*, Fig. 52). Another reason for this difference may be the letter that occurs before the *m*. In Fig. 3, middle column, it appears that the *m* is always ligatured from the preceding letter to its middle part when the letter is a *h*, a *ḥ*, or a *y*.

As for the mixture of both (form 6, right column in Fig. 3), of which there are nine examples, it appears

[10] Twenty examples of form 1 against fifty of form 2.

[11] *CIS* ii 963 (Wādī Mukattab, AD 206), *CIS* ii 2666 (Jabal Munayjah, AD 218–219) and NDGS 2 (Wādī Maghārah, AD 266), *CIS* ii 1491 (Wādī Fayrān, AD 232).

[12] UJadh 360 was considered as "transitional" because of the general form of the name *mᶜnw* (ligatures, the form of the ᶜ) but in fact, none of the letters is really diagnostic.

Stem at the bottom on each side (form 4)			Stem in the middle on each side (form 5)			Stem at the bottom on the right and in the middle on the left (form 6)		
Letter before	*Letter after*	*no.*	*Letter before*	*Letter after*	*no.*	*Letter before*	*Letter after*	*no*
						{b/n}	{b/n}	1
			h	w	2	h	w	1
			ḥ	y	1			
			ḥ	d	1*			
			y	w	1	y	w	2
			y	w	1	y	n	1
l	w	1	{l/n}	w	1	l	w	1
			l	w	2			
l	ʿ	?**						
{b / n}	w	1						
n	w	4	n	w	2	n	w	2
ʿ	ʾ	1						
ʿ	y	1	ʿ	y	1			
ʿ	r	4	ʿ	r	2			
š	w	1				š	ʾ	1
š	ʿ	2	š	ʿ	2			
š	š	1						
t	n	1						

Note that "no." in the headings refers to the number of examples.
* This text (Ḥijr 1) is in fact early Arabic (see Fig. 39).
** There is some doubt about reading this letter as a *m*. It could also be a *q*.

Figure 3. *Letters on each side of the m in forms 4, 5, and 6.*

that it is very much linked to the presence of a *w* after the *m*. Indeed, in six examples out of the nine, the *m* is followed by a *w*, as in ARNA.Nab 17 (Fig. 21).

Ligature on one side only: for this category, one should distinguish the words which start with *m* or end with *m* from those in which the *m* is in medial position. When the word starts with *m*, it is naturally not ligatured from the right and when it ends with *m*, it is of course not ligatured to the left. For the words in which the *m* is in medial position, the presence or absence of a ligature from the right depends very much on the letter which comes before. The database contains fourteen examples, in seven of which the letter that precedes the *m* is an ʾ, a *d*, or a *r*, which are normally not ligatured to the left (see UJadh 3, Fig. 32). In four examples, the unligatured character of the *m* is shared by some of the other letters in the text, which would make the case of the *m* not significant (see, for example, the *n-m* in UJadh 219, Fig. 41, where, however, M.C.A. Macdonald notes that apart from the combinations of letters which quite often lack ligatures — such as *k-y* in *dkyr* — only the letters *s-p* in *ywsp* and *b-ṭ* in *bṭb* are not ligatured). Finally, there are two examples in which we would expect the *m* to be ligatured from the right, after a *y* and a *ḥ*, but where it is not. Note that there is only one example in which the *m* in medial position is not ligatured to the left, possibly Ulā-JSNab 386, but the text is known only from a squeeze (the word in question is *šmʿwn*, l. 3, but M.C.A. Macdonald notes that upright ʿ often does not take a ligature from the right) (see Macdonald 2009*a*: 208 and n. 5). It is clear, therefore, that the letter *m* is normally ligatured to the left, whatever the letter that comes after it.

As for the *position* of the ligature, there are three cases, represented by forms 7–9. In most examples, the stem of the ligature is halfway up the letter: there are twenty-three examples of form 7 (where *m* is followed by ʾ, *h*, *w*, *ḥ*, *k*, *l*, *n*, ʿ, *r*, *š*, or *t*) against three of form 8 and only one of form 9.

Finally, it should be noted that all examples of final *m* in this corpus of texts are derived from the "classical" Nabataean form of the letter, i.e. forms 1 or 2. In the existing examples, however, one should be careful to treat separately the final *m* of *šlm*, a word that may have become, in late texts, an ideogram. In the following examples, the final *m* occurs in words other than *šlm*:

- Ar 19 (Fig. 20): the final *m* in *grʿm* is "classical" whereas the medial *m* in *ʿmrw* is "evolved";
- UJadh 309 (Fig. 48): compare the final *m* in *ywm* (l. 4) and the initial *m* in *mʾt* (l. 6);
- M 1 (Fig. 27): compare the final *m* in *ywm* (l. 3) with the examples of initial and medial *m* in the text (*rm{n}h*, *mn*, *mytt*, *mʾh*, *ḥmš*).

— *p* (Fig. 2). Only two different forms of the letter have been distinguished, the first of which is the "classical" Nabataean form. In our corpus, only twenty texts contain the letter *p*. Among them, seventeen have form 1 and only three have form 2. The clearest examples of form 2 are in the two names in UJadh 222 (Fig. 42). Note that the final form of the *p* of *ywsp* in UJadh 219 (Fig. 41) is almost that found in early Arabic inscriptions. It is very similar to the final *f* of the patronym of the author of the Jabal Says graffito, whose name is *mʿrf*, as suggested by M.C.A. Macdonald in his recent rereading of the text, as well as to the *p* of *ywsp* in the newly published inscription from Taymāʾ (Macdonald, 2009*b*; al-Najem & Macdonald 2009: 210).

— *ṣ* (Fig. 2). There are basically two forms of the letter. The first one is the "classical" form while the second one can be considered as the evolved form. Two other forms are attested in one text each and may be considered as oddities (they are indicated as such on Fig. 2). In one of them, S 3 (Fig. 30), the reading of the letter in the first name, *byṣw*, is not certain. As for the second odd form, it is attested in S 1, which is due to be published by Kh. al-Muaikil (al-Muʿayqil). It occurs in line 1, in the word *ʾṣḥbh*, where it cannot be read as a *š*. Form 2 of the letter occurs in four texts only, one example of which is UJadh 248 (Fig. 43). This form of the letter is very close to the *s* of *ywsp* in UJadh 219 (see Fig. 41) but it cannot be read as a *s* because, as noted by M.C.A. Macdonald, it is ligatured to the left, whereas *s* is not. There is no example of final *ṣ*.

— *q* (Fig. 2). There are also two forms of this letter. Again, 1 is the form in "classical" Nabataean while form 2 has undergone the same development as the letter *p*, becoming a circle on the line in initial/medial position. The developed form of these two letters tends to look very much like the *m* (compare them with form 4 of the *m*). In all our texts except one, the *q* belongs to form 1 but the only example of form 2 is particularly interesting. It occurs in S 3 (Fig. 30), in the name *mrʾlqyš*, where there is no doubt about the reading.

— *r* (Fig. 4). The letter *r* is difficult to analyse because of the great number of small variations. However, I have identified four forms of the letter in the texts. The evolved forms are 2 and 3. Form 4 is an odd form of the *r* usually found in combination with a *b*, which has exactly the same shape, thus forming the word *br* with two inclined parallel lines joined at the base. It is found in only one text of our corpus (unfortunately not among the selected texts presented in Appendix 2) but it occurs in "classical" Nabataean texts from Umm Jadhāyidh, for example in ThNUJ 114, where the letters *bdʾ* in *ʿbdʾlgʾ* are also written as three diagonal strokes. The *r* is one of the letters for which we have the greatest number of examples. There are approximately 100 of them, and they are more less evenly distributed between forms 1, 2, and 3

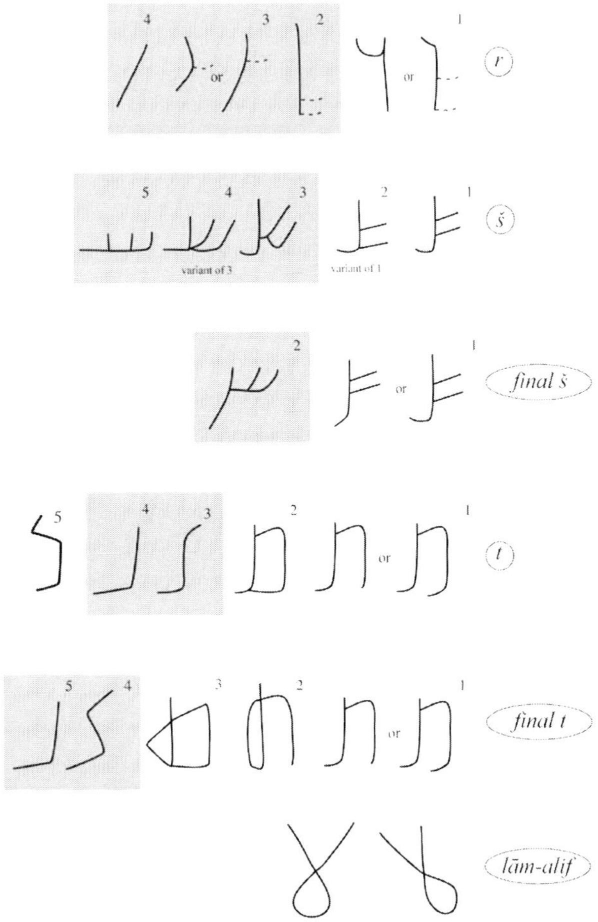

FIGURE 4. *The forms of* r, *medial and final* š, *medial and final* t *and* lām-alif.

of the letter. Very good examples of form 2 can be seen in UJadh 3 (Fig. 32) in *dkyr* and *grmw,* as well as in UJadh 219 (Fig. 41). The variant of form 3, seen for example in S 3 (Fig. 30) and UJadh 222 (Fig. 42), in the word *dkyr*, may be considered as the most evolved form of the letter.

— initial and medial *š* (Fig. 4). The development of medial *š* is very interesting because one can clearly see how the three stems of which the letter is composed move downwards until they become three small vertical strokes resting on a horizontal line, exactly like the Arabic letter. Forms 1 and 2 can be considered as variants of the same form and the same is true of forms 3 and 4 while form 5 is the final outcome of the development. Fifty-two texts contain a medial *š*, out of which the examples in twenty-five are of form 1, i.e. the "classical" Nabataean form.[13]

[13] Form 2 occurs in a few texts such as M 1 (Fig. 27) and the Stiehl inscription (Fig. 31).

Form 3, which is best illustrated in UJadh 309 (Fig. 48) in the words *šly*, *ʾwšw*, *šlm*, and *šnt*, occurs in eight texts. Note that in UJadh 309 the *š* of *tšʿyn* is clearly on its way to form 5. Other examples of form 3 can be seen in JSNab 18 (in *šlm*, Fig. 25), UJadh 266 (in *šlm*, Fig. 44), and M 1 (in *šnt*, Fig. 27). Finally, form 5 can be best seen in UJadh 266 (in *ʿšylh*, Fig. 44) and in UJadh 299 (Fig. 47).

— final *š* (Fig. 4), of which there are only eight examples, appears in two forms, the second of which can be considered as the "evolved" one. The best example is the name *ʿbdʾyš* in UJadh 105 (Fig. 37). The evolution of the final form 2 from form 1 is understandable only if we assume that there was an intermediate form equivalent to form 3 of the medial/initial sequence. However, this form does not occur in our texts.

— initial and medial *t* (Fig. 4). All the texts that have form 1 of medial *t*, which is the normal form in "classical" Nabataean, are either third-century texts (*CIS* ii 963, 1491, NDGS 2, *RES* 528, B 3, JSNab 17) or formal ones such as Stiehl and al-Namārah. UJadh 309 (Fig. 48), which is dated to AD 295, also has this type of *t*. Finally, *CIS* ii 333 has it also but the text is not clearly dated. The *t* which is closed at the bottom (form 2) is a variant of form 1 and occurs in two texts only: Stiehl (Fig. 31) and ARNA.Nab 17 (Fig. 21). M.C.A. Macdonald notes that it is not simply the base line being continued under the *t*, but that even isolated examples have a closed base in the Stiehl inscription. It is not clear how forms 1 and 2 evolved to forms 3 and 4 because intermediate stages are missing. According to A. Yardeni (2000: 263), however, an intermediate phase in the evolution of the cursive *t* may be reconstructed between the looped form of the letter (which appears in the papyri not only in final but also in medial positions) and the form resembling the early Nabataean *y*. This reconstruction is described as follows: in the intermediate phase, the right stroke would become longer than the left one, the loop would gradually close until a mere angle remains between the strokes. This angle would then gradually open up, leaving only a wavy stroke. UJadh 297 (Fig. 45) has a very good example of form 3. Note that form 5 is attested only in LPNab 41 (Fig. 26), in the word *tnwḥ*.

— final *t* (Fig. 4) can have five forms. The first is the "classical" Nabataean form and is identical to the first initial/medial form. Form 2 (and 3, which is a variant of 2) is a particular form of final *t* which is known also in "classical" Nabataean texts and which is best seen, in our corpus, in JSNab 17 (*ḥrtt*, *brt*, *hlkt*, *šnt*, Fig. 24), in Stiehl (*brt*, *mytt*, *šnt*, Fig. 31), *CIS* ii 963 (*šnt*, *tltt*, Fig. 23), etc. As was the case for the medial form, we lack the intermediate forms between forms 1–3 and forms 4–5 (but see A. Yardeni's comment quoted above). Form 4 is well represented in our corpus, as in *šnt* in UJadh 109 (= ThNUJ 132–133, Fig. 38) and UJadh 297 (Fig. 45). Form 5 can be seen only in UJadh 298 (in *ʾlḥrt*, Fig. 46), where it may however, as suggested by M.C.A. Macdonald, be simply a lazy version of form 4 rather than a separate form (see the possible slight curve to the right near the top of the letter). In UJadh 309, one can see that the ligatured form of the letter (in *šnt*) is form 4 while the unligatured form (in *mʾt*) is form 3 (Fig. 48).

— *lām-ʾalif* (Fig. 4). In the texts in transitional script, this combination of letters usually receives special treatment, as already in the Namārah inscription. They are combined in such a way that they seem to form one single letter. UJadh 367 (Fig. 51) gives extremely clear examples of *lām-alif* preceded by *ʾ* in the name *ʿbdʾlʾšḥn*, repeated twice. The *lām-alif* there clearly has the form found in the Namārah inscription as well as in the Arabic inscriptions of the first century AH.

The corpus of transitional texts

The corpus of transitional texts I have collected so far contains 116 texts, one of which may have to be removed because it could be considered as "classical" Nabataean (UJadh 360, Fig. 50, see above). One other text, Ḥijr 1 (Fig. 5), which reads *ʿly mḥmd*, was initially thought to be transitional but is in fact Arabic. M.C.A. Macdonald notes that the forms of all the letters, including the *d*, are perfect early Kufic.[14] However, with the exception of the *d* and the *y*, the form of the letters would also fit relatively well with the last stages of the development of the Nabataean script.[15]

Some of these 116 texts are published and have been known to the academic community for a long time. This is the case of the inscriptions from Sinai published in the *CIS* ii and by A. Negev (see the table below), as well as the Namārah inscription, JSNab 17 and JSNab 18, Stiehl, etc.

Some others have been published more recently, mostly in Arabic, by S. al-Theeb (al-Dhīyīb),

[14] See for instance the inscription on plates of copper at the entrance to the Dome of the Rock (Grohmann 1971: pl. 12/a), where the final *y* of *al-ḥusnā* in line 5 and *nunajjī* in line 6 has this form (as opposed to the final *y* in other words, e.g. *ʿalā*), and see also the *d* in *Muḥammad* in line 7.

[15] As noted by R. Hoyland, the *y* comes more directly down than in any of the Nabataean texts (where it tends to sweep to the left side first before descending). R. Hoyland also points to an effortlessness and smoothness in the text, which does not exist in the other inscriptions.

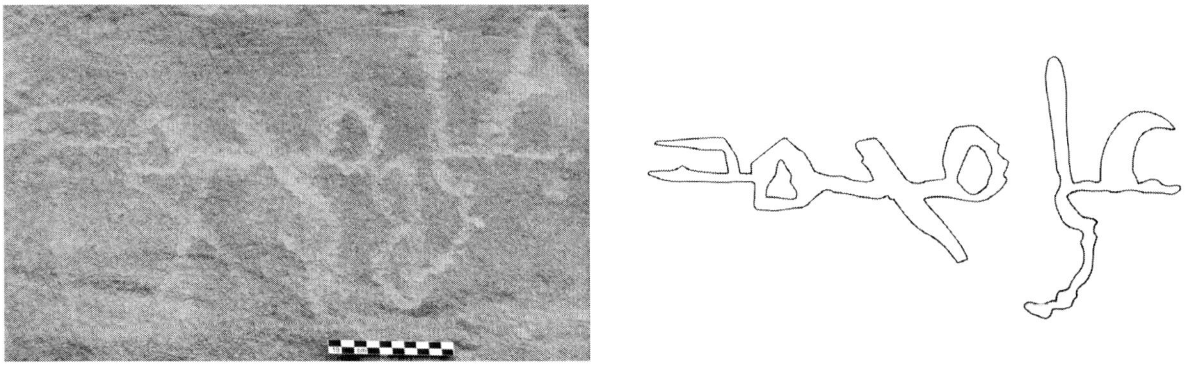

FIGURE 5. *Ḥijr 1. (Photograph Madāʾin Ṣāliḥ archaeological project, facsimile L. Nehmé).*

Kh. al-Muaikil (al-Muʿayqil), M. al-Muraykhī, and ʿA. al-Ghabbān or in English, by M.C.A. Macdonald (see the bibliography under each of these names).

Some are miscellaneous texts collected by scholars who were given photographs taken either during surveys or haphazardly. This is the case, for instance, of the photographs taken by F.V. Winnett and W.L. Reed in 1962, some of which show texts that are relevant to our corpus.

Finally, the majority are texts that were recorded during the survey of the Darb al-Bakrah, which is the name given to the ancient caravan road between Ḥegrā and Petra, named after a pass in the mountains south of Tabūk. This survey was directed by ʿA. al-Ghabbān, and the present author is in charge of the publication of the texts in the Nabataean script, which were photographed in 2004.

It should also be noted that fifty of these 116 texts were examined during a workshop organized by the present author in Paris in 2005 as part of a project on the development of the Nabataean script. During this workshop, seventy-one texts dating from the beginning of the third century AD to the end of the reign of the Umayyad caliph Muʿāwiyah, in AD 680, were read by the participants who were, apart from the author, ʿAlī al-Ghabbān, Robert Hoyland, Michael Macdonald, Khalīl al-Muʿayqil, Mshallaḥ al-Muraykhī, Christian Robin, and Moulay Janīf. At the end of the workshop, the intention was to publish the whole corpus of texts dated to this time span. However, the publication was delayed and, in the meantime, the collection of texts examined during the workshop has been rendered incomplete by the discovery — mainly during the Darb al-Bakrah Survey — of dozens of new texts, most of which were previously unknown.

Two regions have provided the greatest number of texts either dated to between the third and fifth centuries AD regardless of the degree of development their script shows, or written in transitional characters: Sinai and north-west Arabia, the latter being the region which, in the last few years, has been found to be the richest in late Nabataean, transitional, and early Arabic texts (Fig. 6).

Dots on letters

The only letter that receives a dot in the transitional texts is the *d*, and this is the case in twenty-eight texts in this corpus (within our selection, see JSNab 17, QN 2, Stiehl, UJadh 3, 105, 109, 178, 248, 375, 405, ʿUlā 1). J. Healey (in Healey & Smith 1989: 78), followed by C. Robin (2006: 364, and fig. 16) has suggested that, in JSNab 17, diacritical dots were placed on two other letters: on the *r* of *ḥrtt* and on the *š* of *rqwš*. In both cases, however, an examination of the original shows that these are chips in the stone and not dots marked intentionally.

Outside our corpus of 116 texts, the use of dots is also not rare and can be found in several inscriptions, where it does not seem to be used exclusively for *d*:[16]

- JSNab 65 (Fig. 8), from the Jabal Ithlib area in Madāʾin Ṣāliḥ. The text reads *dkyr lwqys* ʾ---- *ʿdrw bṭb* and there is a dot on the *d* of *ʿdrw* which was not recorded by Jaussen and Savignac;
- JSNab 123 (Fig. 9), also from the Jabal Ithlib area. The dot on the *d* of *dkyr*, at the beginning of the text, was not recorded by Jaussen and Savignac;
- JSNab 181 = *CIS* ii 320D (Fig. 10) from Mabrak an-Nāqah north of Madāʾin Ṣāliḥ. The text reads *ʿylw br šw{d/r}{d/r}mʾ šlm*, and the fourth letter of the patronym has a dot over it. It was read Šūdrūmā by Jaussen and Savignac, thus suggesting that the

[16] Some of these references had already been collected by M.C.A. Macdonald, to whom I am very grateful for making them available to me. On the use of diacritics in Nabataean, see Healey 1990–1991: 45.

Site name	*Region*	*No. of texts*	*References*
al-Aqraᶜ	North-west Arabia	1	– al-Muraykhī & al-Ghabbān 2001
al-ᶜArniyyāt	North-west Arabia	4	– four unpublished texts, one of which is presented in Appendix 2 under "**Ar 19**"
ᶜAvdat	Negev	1	– *RES* 528
Boṣrā	Ḥawrān	1	– **B 3** (Unpublished. Note that this text is written in what M.C.A. Macdonald [2003: 44–46, 54–56] calls "Ḥawrān Aramaic")
Dūmat al-Jandal	North-west Arabia	1	– **ARNA.Nab 17** (= Macdonald 2009*a*)
Madāᵓin Ṣāliḥ and vicinity	North-west Arabia	9	– **JSNab 17–18** (see Appendix 2, *s.v.*), – **Stiehl** (= Stiehl 1970), and – six unpublished texts, including Ḥijr 1 (Fig. 5) which is in fact Arabic
Jabal Munayjah	Sinai	1	– *CIS* ii 2666
Jubbah	North-west Arabia	1	– *CIS* ii 345
Mābiyāt	North-west Arabia	1	– **M 1** (= al-Muraykhī, in press)
al-Namārah	Ḥawrān	1	– see Calvet & Robin 1997: 265–269, no. 205
Qāᶜ al-Nuqayb	North-west Arabia	1	– **QN** 2 (Unpublished)
Ṣadr Ḥawẓāᵓ	North-west Arabia	1	– One unpublished text
Sakākā (Qalᶜah)	North-west Arabia	10	–Two texts, one of which is S 3 (= al-Muaikil 2002 [=1993] unpublished in Winnett & Reed 1970 but it appears in one of their photographs, reproduced in Macdonald 2009a: pl. 4, no. 13a) – al-Muaikil & al-Theeb 1996: nos 30 (= ARNA.Nab 10) and 31 (= ARNA.Nab 13) – ARNA.Nab 2, 6–9, 11–12 (for these texts, see also Macdonald 2009a: pl. 4. [no. 7 = al-Theeb 1994a: 190, no. 6])
Sakākā (vicinity)	North-west Arabia	4	– al-Theeb 2005: no. 3 – **S 1** (= al-Muaikil, forthcoming) – Two unpublished texts
Taymāᵓ	North-west Arabia	1	– al-Najem & Macdonald 2009
al-ᶜUdhayb	North-west Arabia	2	– Two unpublished texts
al-ᶜUlā and vicinity	North-west Arabia	3	– JSNab 386, – *CIS* ii 333 – **ᶜUlā 1** (Unpublished)
Umm al-Jimāl	Southern Ḥawrān	2	– **LPNab 41** – LPArab 1
Umm Jadhāyidh	North-west Arabia	61	– al-Theeb 2002[17] + 46 unpublished texts, twelve of which are presented in Appendix 2, under the siglum **UJadh**
Wādī Fayrān	Sinai	1	– *CIS* ii 1491
Wādī Ḥajjāj	Sinai	1	– Negev 1981: no. 9
Wādī Maghārah	Sinai	2	– NDGS 1–2
Wādī Mukattab	Sinai	1	– ***CIS* ii 963**
Wādī Ramm	Ḥismā	6	– Grimme 1936: 90–95 (= Gruendler 1993: 13, A1)[18] – Four unpublished texts

[17] See inscriptions nos 14, 30, 34, 38, 41, 45, 48, 62, 84, 122, 128, 132–133, 145, 159, 203.

[18] See Hoyland (this volume), who identifies two texts.

Figure 6. *Table showing the provenances of the inscriptions with transitional letter forms and texts dated to the period AD 200–500. Those for which readings are given in Appendix 2 are in bold script.*

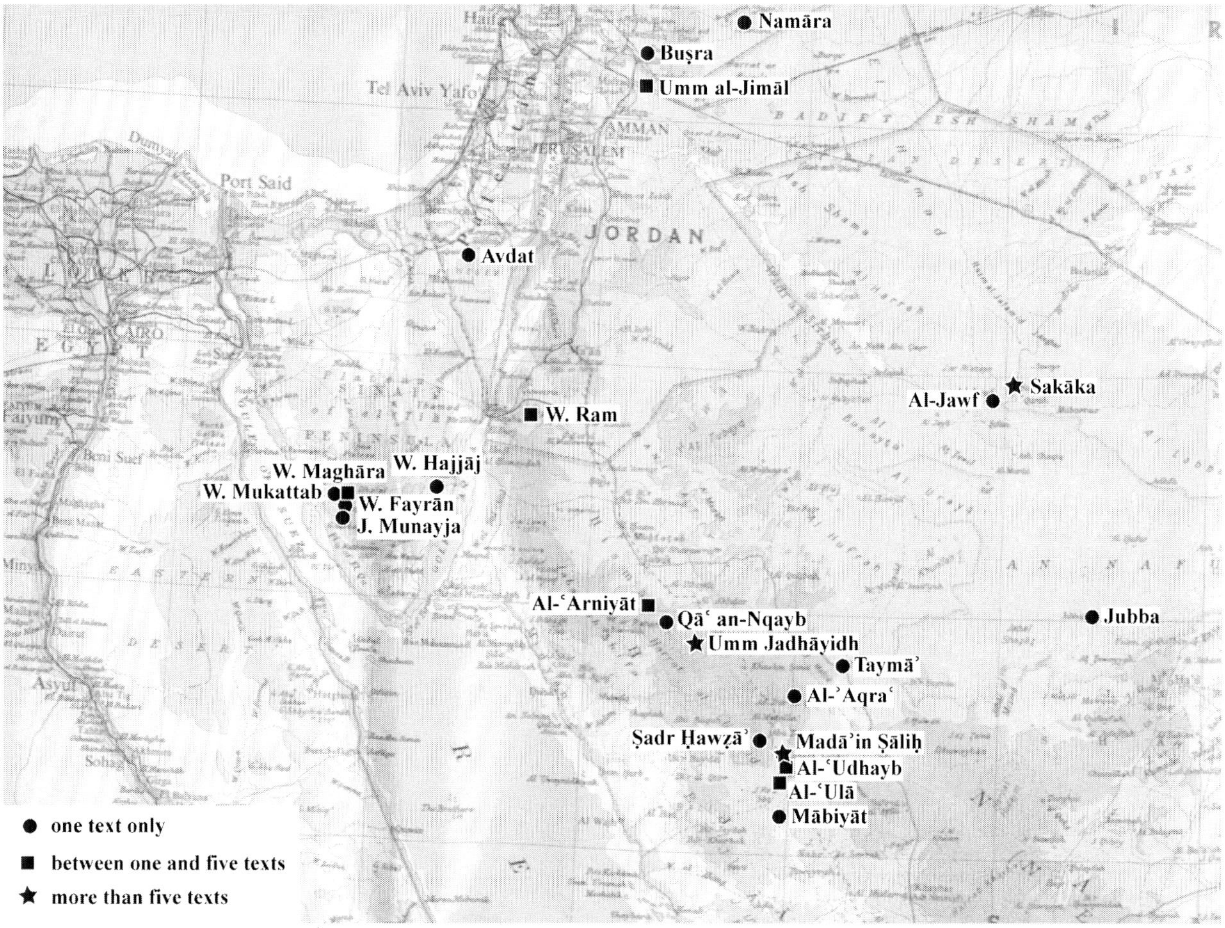

Figure 7. *The geographical distribution of the texts in the transitional script.*

dot was carved above a *r*, but it may well be read Šūrdūmā, with a dot on the *d*. Neither of the names is attested in the Nabataean corpus but *s*^{2}*rḏm* occurs once in Safaitic, in *CIS* v 1955;

- JSNab 212, from around the railway station of al-ʿUlā. The text is known only from a copy in which a dot appears above the *r* of *br*. Its presence, should, however, be checked on the original;
- JSNab 321 from Sheqeiq edh-Dhib (Shuqayq al-Dhiʾb), *c.*20 km north-west of Madāʾin Ṣāliḥ. The text is known only from a copy. It reads *šlm bgrt br brdw bšnt 36 lrbʾl*[19] and there is a dot above the *d* of *brdw*;
- *CIS* ii 344, from the area of Taymāʾ. The text is known only from J. Euting's copy. It reads *ʿ{b}dt br t{y}mw br prk* and the dot is above the *d* of *ʿbdt*;
- ThNUJ 81, from Umm Jadhāyidh (Fig. 11). The text reads *dkyr wbrw br ʿdrw bṭb* and there is a dot on both examples of *d*;
- an unpublished text from Madāʾin Ṣāliḥ which is carved 1 m to the left of the eye-betyl which is reproduced in Jaussen and Savignac[20] (Fig. 12). This text is best read *šlm ddn*. There is a dot on each example of *d*;
- an unpublished text from the Darb al-Bakrah Survey, Ir 3, from the site of ʿĪrīn (Fig. 13), which reads *šlm rbybw br mšlmw* and in which both examples of *r* have a dot above them;

[19] *Contra* Jaussen and Savignac, who read the patronym as Nadrū and suggest therefore that the dot is above a *r* rather than a *d*. This rereading of the patronym was suggested by Macdonald (2009*a*: n. 51).

[20] Jaussen & Savignac 1909–1914, i: 426 and fig. 217 (it bears the number Ith 55 in the new catalogue of the monuments of Madāʾin Ṣāliḥ). This text was photographed by J. Bowsher and identified by M.C.A. Macdonald as having diacritical dots. It is part of the epigraphic point no. 83, associated with the eye-betyl, which contains the inscriptions JSNab 111–118 as well as twenty-six unpublished texts.

Figure 8. *JSNab 65. (Photograph Madāʾin Ṣāliḥ archaeological project).*

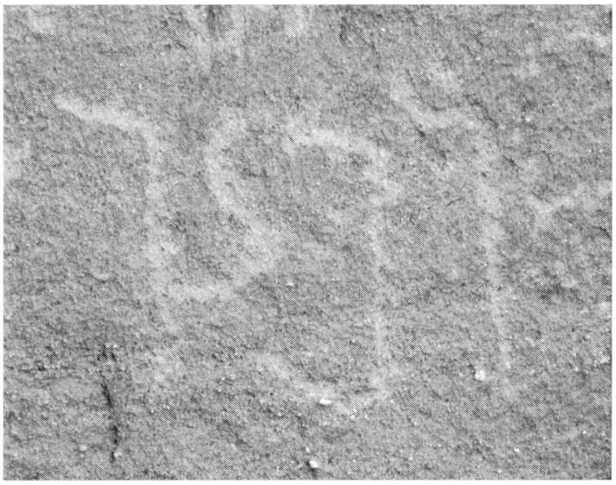

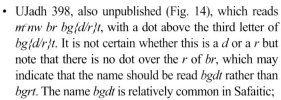

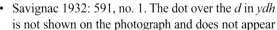

Figure 9. *JSNab 123. (Photograph Madāʾin Ṣāliḥ archaeological project).*

Figure 10. *JSNab 181. (Photograph Madāʾin Ṣāliḥ archaeological project).*

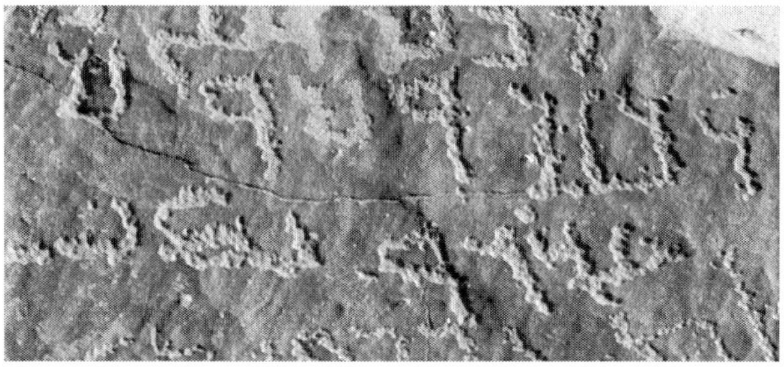

Figure 11. *ThNUJ 81. (Photograph Darb al-Bakrah project).*

Figure 12. *Unpublished text from Madāʾin Ṣāliḥ (epigraphic point no. 83). (Photograph Madāʾin Ṣāliḥ archaeological project).*

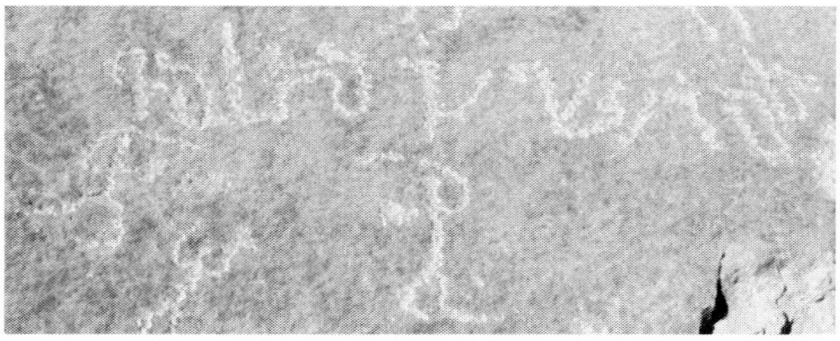

Figure 13. *Ir 3. (Photograph Darb al-Bakrah project).*

- UJadh 398, also unpublished (Fig. 14), which reads *mʿnw br bg{d/r}t*, with a dot above the third letter of *bg{d/r}t*. It is not certain whether this is a *d* or a *r* but note that there is no dot over the *r* of *br*, which may indicate that the name should be read *bgdt* rather than *bgrt*. The name *bgdt* is relatively common in Safaitic;
- Savignac 1932: 591, no. 1. The dot over the *d* in *ydh* is not shown on the photograph and does not appear on the squeeze but it was noticed by G.M.H. King on the stone and appears on the photograph she took;
- al-Theeb 1994*b*, inscription B: 36–38. There is a dot over *d* in *ḥdh* but not on the *d* of *ʾšdw*, which should perhaps be read *ʾšrw*.

Thus in most cases, the dot is used in Nabataean to distinguish the *d* from the *r* and is generally put on the *d*. There are only two texts (JSNab 212 and Ir 3) in which it

Figure 14. *UJadh 398. (Photograph Darb al-Bakrah project).*

Figure 15. *UJadh 118–119. (Photograph Darb al-Bakrah project).*

is clear that the dot is above the *r*. In Arabic, the earliest examples of texts using diacritical marks are the Ahnās papyrus (Grohmann 1932: 32–34 [not seen, quoted in Robin 2006: 342]; see also Larcher, this volume: fig. 6) and the Zuhayr inscription (al-Ghabbān 2008), dated respectively to 22 and 24 AH, to which should perhaps be added the inscription written on a piece of wood found in Petra among the Greek papyri, published by O. al-Ghul in 2004.[21] As shown by C. Robin, who recently re-examined the use of diacritics (2006: 343–345), eleven letters out of the fifteen which, from the third century AH onwards, bear diacritics, already appear with them in the early documents dated up to the reign of Muᶜāwiyah. However, neither the *d* nor the *r* are among these letters. This is to be expected in Arabic because these two letters are no longer homomorphs.[22] It is interesting to note that in the transitional texts, which are supposed to be "on their way to Arabic", none of the letters which will later receive diacritics receive any and, surprisingly, the *d* is still almost the only letter which receives a dot, despite the fact that its form is clearly distinct from that of the *r*. Thus, among the twenty-eight texts of our corpus which bear a dot above the *d*, twenty-four contain also a *r* which has a form which is very distinct from that of the *d*.[23] In these twenty-four texts, therefore, there was no *need* to put a dot above the letter *d* to distinguish it from the *r*. All in all, the habit of adding a dot on the *d* in "classical" Nabataean in order to distinguish it from the *r*, which had a very similar shape, was rare but nevertheless existed and there is a possibility that this use in later texts was inherited from "classical" Nabataean. Strangely enough, the use of the dot to distinguish the letters which were starting to become homomorphs (such as the *d* and the *k*, the *n* and the *b*, etc.) is attested neither in the corpus of texts we collected nor in any of the pre-Islamic Arabic texts of the sixth century AD.

The relationship between the "classical" Nabataean and transitional script

My intention, in this section, is not to show similarities or dissimilarities between the scripts in these two categories of texts but to present a few examples of inscriptions in the two scripts whose spatial relationship — including superposition — allows us to establish a relative chronology between them.

The first example is UJadh 118 (= ThNUJ 122) and UJadh 119 (= ThNUJ 123) (see Fig. 15). UJadh 118 is a transitional text while UJadh 119 is a very "classical" Nabataean one and it is clear that the tail of the final letter

[21] M.C.A Macdonald (2008: 467a) questions the authenticity of this piece for the following reasons: "the relatively late form of the final letter, the arrangement of the diacritical dots under the *yā*ʾ, and the difficulty of interpreting the word as anything other than the modern name *Nāyif*".

[22] For this term see Macdonald 1986: 148, n. 119.

[23] Ar 34, Ḥijr *2*, *3*, 4, 5, ***Stiehl***, **JSNab 17**, ***QN 2***, UJadh *3*, *4*, *11*, *67*, *69*, ***105***, ***109***, *118*, *122*, ***178***, *220*, ***248***, *287*, *300*, *301*, 312, *343*, ***375***, *386*, ***405***, *467*, ***ᶜUlā 1***. The numbers in bold are presented in Appendix 2 and those in italics contain *d* and *r* with distinct forms.

Figure 16. *UJadh 27 and 31. (Photograph Darb al-Bakrah project).*

Figure 17. *UJadh 343–344. (Photograph Darb al-Bakrah project).*

of UJadh 118 runs across and *over* line 1 of UJadh 119. The Nabataean text reads *šlm ḥny br krys bṭb / w šlmw ʾḥwhy*; the transitional one reads *dkyr z{b/n}y{b/n}w / bṭy*, with a dot on the *d* of *dkyr*.

The second example is UJadh 27, written in "classical" Nabataean, and UJadh 31, written in transitional characters (Fig. 16). UJadh 27 reads *ʿbdʿbdt br ʾbw* while UJadh 31 reads *dkyr ʾpṭy / br ʾwšw bṭb / w šlm*. The bottom of the stem of the second *w* of *ʾwšw* of UJadh 31 clearly runs over the top of the *d* of UJadh 27 and must therefore have been written after it.

The third example, represented by UJadh 343 and 344 (Fig. 17), is particularly interesting. The text in "classical" Nabataean characters (UJadh 344) reads *zpr br yʿmr / w ʿbydw šlm*, and the text in transitional characters (UJadh 343), reads *dkyr lʿmrw / br zbʾbrh bṭb / w šlm*. Parts of two of the letters of the "classical" Nabataean text run *over* two letters of the text in transitional characters. This can be seen firstly at the top of the *y* of *yʿmr* of UJadh 344, which must have been carved *after* the *b* of *zbʾbrh* of UJadh 343;[24] and secondly at the top of the *r* of *yʿmr* which, again, must have been carved *after* the *b* of *bṭb*. This order in the superposition of the characters shows either that the letter forms of the "classical" Nabataean script continued to be used along with the transitional forms of the same letters or that the transitional forms of the letters started to be used early.

Concerning the use of the "classical" Nabataean script at a late period, note should be made of the inscription UJadh 172 (Fig. 39) which, if it is indeed dated to AD 311–312 rather than AD 151–152, may be a late attestation of this script.

The letter forms attested in the selection of texts presented in Appendix 2

Commentary

Only two letters appear in both their "classical" and transitional forms in the same inscription: the *ʾ* and the *m*. In JSNab 18, the *ʾ* in *ktbʾ*, line 2 (Fig. 25) cannot be form 1 because there has to be room for the *b* before it. One has therefore to assume that the sign which follows the *t* is a combination of the *b* and the *ʾ* (a similar combination can also be found in UJadh 343, Fig. 17), the latter being an evolved form of the letter (form 2 on Fig. 1). All the other examples of *ʾ* in this text, however, belong to form 1. In M 1 (Fig. 27), the *ʾ* in *bʾyr* and *mʾt* in line 4 belongs to the straight form of the *ʾ*, in contrast to the looped form in *ʾntth* in line 1 and *qryʾ* in line 2. In UJadh 10 (Fig. 33), there are two examples of *m* which have a transitional form, in *šlymw* and *ʿmrw*, and one which has a "classical" form, in *šmnw*, in none of which is *m* in final position (see the discussion of the letter *m* above).

The evolved form of the *ʾ* does not appear in inscriptions before the last quarter of the third century, in combination with a *b*, in *ktbʾ* and *bʾyr*, respectively in JSNab 18 (Fig. 6) and M 1 (Fig. 27), and isolated in ARNA.Nab 17 (Fig. 21). It is then much more regularly and often used in the texts than the "classical" form. However, forms 2 and 3 already appear in the Nabataean papyri along with form 1 (Yardeni 2000: 237).

The form of the *g* (no. 1 on Fig. 1), which looks like the evolved *ḥ* (no. 3 on Fig. 1), is less widespread in this corpus of texts than the "evolved" form of the *g*. In the

[24] The letter that follows the *z* looks like an early Arabic *ʾ* but if we compare the shape of the letter with the signs for *bʾ* in *ktbʾ* as they appear in JSNab 18 (see below), it is almost certain that the letter after *z* should be read *bʾ*.

whole corpus of transitional texts, there are only four examples of the former and ten examples of the latter, out of which four seem to have more or less completely lost the lower part of the diagonal stroke. M.C.A. Macdonald has pointed out to me that this is interesting in view of the fact that this bottom part of the diagonal is characteristic of early Arabic *g*, and that *g* and *ḥ* are homomorphs from the earliest examples of the Arabic script.

The *d* and the *k* can be treated together. The form of the *d* in the transitional texts is characteristic and appears at the end of the third century in ARNA.Nab 17 (Fig. 21), but both forms 1 and 2 continue to be used in a relatively large number of texts, both dated and undated, whereas in these texts the *k* has form 2 much more regularly than form 1. A. Yardeni has pointed out that the form of the *d* we have in the graffiti from the third and fourth centuries AD is not a late development of the letter because it resembles almost exactly that in earlier Aramaic texts (2000: 241). M.C.A. Macdonald notes however that this does not necessarily mean that it is an "inheritance" from earlier Aramaic since it may have developed this form from the "classical" Nabataean form.

Medial *h*, along with *ḥ*, is one of the letters which evolves early in the development of the script. Form 2 of this letter appears in JSNab 17 (Fig. 24)[25] and is used throughout the texts. It can thus be considered as a diagnostic letter. Form 2 of the letter is not widespread in the papyri.

Evolved forms of final *h* (forms 2 to 4) appear for the first time in an early third-century text from Sinai, *CIS* ii 963 (Fig. 23), dated (with some uncertainty) to AD 205–206. From AD 280 onwards (in M 1 [Fig. 27] and others), they are almost always used in the texts, except in JSNab 17 (Fig. 24), LPNab 41 (Fig. 26), and Stiehl (Fig. 31), which are formal texts. In the papyri, both the long final *h* (in what A. Yardeni calls "calligraphic" cursive script) and evolved forms (in what she calls "extreme" cursive script, i.e. the result of rapid writing) are used (Yardeni 2000: 242).

Within form 1 of *w* in Fig. 1, I have distinguished, on the one hand those cases where the letter is ligatured from the preceding letter (in which case the ligature is clearly made to the bottom of the *w*), and on the other, those in which the *w* is *not* ligatured from the preceding letter. It appears that form 2, in which the ligature is made to the middle part of the *w*, more or less at the level of the loop, is much more widespread in the texts in this corpus than the ligature to the base, as in form 1 (twenty-one against five). In the papyri, the *w* keeps its loop in the "calligraphic" cursive script whereas in the "extreme" cursive script, the loop is rendered by a mere thickening at the top of the letter (Yardeni 2000: 244).

The *ḥ* is, like the *h*, a letter that starts evolving early and is more diagnostic than others. The transitional form is used almost exclusively in all the texts of our corpus, except in LPNab 41 (Fig. 26), in JSNab 17 (Fig. 24, in *byrḥ* only, where an odd version of the "classical" form is used, while all the other examples of *ḥ* in the text are evolved), and in UJadh 19 (Fig. 35) which, again, has a corrupted "classical" form. In the "calligraphic" cursive script of the papyri, the *ḥ* is very close to form 2 whereas in the "extreme" cursive script, it is like form 3 (Yardeni 2000: 246).

Both the medial and final forms of *y* are also very distinctive. The "classical" medial form (no. 1), which was also used in final positions in the first century AD, is no longer used at all. As for medial *y*, the only exceptions to the use of form 2 are an early text (*CIS* ii 963, AD 206–206, Fig. 23), and UJadh 172 (Fig. 39), the date of which is doubtful (either AD 151–152 or AD 311–312). In the papyri, the medial *y* has mostly the wavy form 2, with small variations, and the final *y* has form 2 (Yardeni 2000: 248–249).

I have already mentioned that only one text mixes the "classical" and the evolved forms of the letter *m* in non-final position, UJadh 10 (Fig. 33). One interesting feature about the *m* is that the final form always retains the "classical" Nabataean shape. If we ignore the row "*m* 1 *f*" in Fig.18, we realise that a large majority of the examples of *m* in this selection of texts (twenty-five against eight) have the evolved forms (4–9), but that the "classical" form goes on being used much more than, for instance, those of *h*, *ḥ*, and *y*. Circular forms of *m* are already very well attested in the papyri (Yardeni 2000: 253).

There is only one clear example[26] of the evolved form of medial *p*, a letter which otherwise has a form which remains very stable throughout the corpus. Evolved form 2 of *p* does not seem to be attested in the papyri. The same is true of the *q*, of which there is only one evolved medial form in our graffiti, in *mrʾlqyš* of S 3 (Fig. 30).

The *r* is one of the letters which varies considerably and it is sometimes difficult to attribute a particular example to one of the forms that have been distinguished. There is, however, a general tendency for the letter to lose its upper bar or flourish and — mainly when combined with *b* in the word *br* — to become smaller and to be ligatured from the preceding letter to its middle part. It

[25] Apart from the texts which are listed in Fig. 18, it also occurs in NDGS 1 from Wādī Maghārah, dated to AD 265–266, and in the Namārah inscription of AD 328.

[26] There are two possible other examples in the rest of the corpus, including in the inscription in al-Muraykhī & al-Ghabbān 2001.

	3rd century								4th c.			5th c.		Undated texts																						
	CIS 963. 205–206?	**B 3. 230–231**	**JSNab 17. 267**	**JSNab 18. 267?**	**ARNA.Nab 17. 275–276**	**M 1. 280**	**UJadh 309. 295**	**LPNab 41. 3rd c.**	**UJadh 297. 305–306**	**UJadh 172. 311–312?**	**Stiehl. 356**	**S 1. 428**	**UJadh 109. 455–456**	Ar 19	QN 2	S 3	UJadh 3	UJadh 10	UJadh 15	UJadh 19	UJadh 90	UJadh 105	UJadh 178	UJadh 219	UJadh 222	UJadh 248	UJadh 266	UJadh 298	UJadh 299	UJadh 360	UJadh 367	UJadh 375	UJadh 405	ʿUlā 1	TOTAL OF x	TOTAL OF x white/grey
ʾ 1	x		x	x		x	x			x	x				x					x															9	9
ʾ 2				x	x							x	x			x													x						6	
ʾ 3						x			x			x										x			x	x		x			x			x	9	15
g 1								x																						x					2	2
g 2					x						x			x			x																		4	4
d 1/2	x		x	x		x		x		x							x			x	x	x		x	x								x	x	14	14
d 3		x			x		x				x	x	x	x	x	x		x	x				x			x					x	x			16	16
h 1	x	x																																	2	2
h 2			x	x				x					x																			x		x	6	6
h *f* 1			x					x			x																								3	3
h *f* 2	x					x																					x				x				4	
h *f* 3						x	x		x			x																x							5	
h *f* 4															x														x						2	11
w 1 ligatured			x		x			x		x																				x					5	5
w 1 unligatured			x				x		x	x	x											x	x	x					x	x					10	10
w 2			x			x	x		x			x	x		x	x	x		x	x	x		x	x	x	x	x		x			x		x	20	
w 3																					x														1	21
ḥ 1								x																											1	
ḥ 2			x																	x															2	3
ḥ 3			x	x		x	x				x	x	x								x							x			x			x	11	11
y 1	x									x																									2	2
y 2	x		x	x	x	x	x	x	x		x	x	x		x	x	x	x	x	x	x	x		x	x	x	x							x	24	
y 3					x	x			x			x	x	x			x		x		x		x						x			x	x		13	37
y *f* 1																																			0	0
y *f* 2	x	x	x	x	x	x	x	x		x	x		x				x								x							x	x		15	15
k 1	x	x	x	x						x														x										x	7	
k 1 *f*								x																											1	8
k 2					x		x		x			x	x	x	x	x	x	x	x		x	x	x			x						x	x		17	17
m 1	x			x						x	x							x												x					6	
m 1 *f*		x		x		x	x					x	x	x					x			x	x	x				x	x			x			14	
m 2																																			0	
m 3			x					x																											2	22
m 4												x		x													x		x						4	
m 5													x							x	x											x	x	x	6	
m 6					x							x							x				x												4	
m 7					x	x	x					x				x	x	x	x					x											9	
m 8						x										x																			2	
m 9																																			0	25

Note: the letters followed by a number in the first column of this table refer to the synthesized letter forms on Figs 1, 2, and 4. The rows with a grey background are the transitional letter forms, and those with a white background are the "classical" Nabataean forms. A letter followed by "*f*" indicates a final form.

FIGURE 18. *Description of characters.*

	3rd century								4th c.			5th c.		Undated texts																						
	CIS 963. 205–206?	**B 3. 230–231**	**JSNab 17. 267**	**JSNab 18. 267?**	**ARNA.Nab 17. 275–276**	**M 1. 280**	**UJadh 309. 295**	**LPNab 41. 3rd c.**	**UJadh 297. 305–306**	**UJadh 172. 311–312?**	**Stiehl. 356**	**S 1. 428**	**UJadh 109. 455–456**	Ar 19	QN 2	S 3	UJadh 3	UJadh 10	UJadh 15	UJadh 19	UJadh 90	UJadh 105	UJadh 178	UJadh 219	UJadh 222	UJadh 248	UJadh 266	UJadh 298	UJadh 299	UJadh 360	UJadh 367	UJadh 375	UJadh 405	ʿUlā 1	TOTAL OF x	TOTAL OF x white/grey
p 1			x					x	x				x											x		x						x			7	
p 1 *f*						x																													1	8
p 2																									x										1	1
ṣ 1			x																																1	1
ṣ 2																										x									1	1
q 1	x		x	x		x									x																				5	5
q 2																x																			1	1
r 1	x		x	x		x		x		x	x	x			x					x					x					x					12	12
r 2					x		x					x	x	x			x	x	x		x	x	x	x	x						x				14	
r 3							x		x			x	x	x		x	x	x			x					x	x	x	x			x	x	x	16	
r 4																																			0	30
š 1	x	x			x	x		x		x	x		x					x	x					x							x				12	
š 2						x					x																								2	14
š 3				x			x																												2	
š 4									x			x										x	x				x								5	
š 5							x					x														x	x	x	x		x	x			8	15
š *f* 1			x			x																													2	2
š *f* 2																x						x													2	2
t 1	x	x	x	x			x				x																								6	
t 2	x										x																								2	8
t 3						x			x			x																							3	
t 4																																			0	
t 5								x																											1	4
t *f* 1										x																									1	
t *f* 2			x					x			x																								3	
t *f* 3							x				x																								2	6
t *f* 4						x	x		x			x	x																						5	
t *f* 5																												x							1	6

FIGURE 18 (continued). *Description of characters.*

should, however, be noted that form 2, a vertical line with the ligature to the bottom of the letter, is also very well represented in the texts.

The *š* is an interesting letter because it is the one in which the evolution of the letter from form 1 to form 5 can be traced through many examples. It should be noted, however, that the "classical" form continues to be used throughout the corpus in more or less the same proportion as the evolved forms (3–5): i.e. fourteen against fifteen. The presence of a "classical" *š* in a text does not, therefore, mean that this text is early. As for the appearance of form 3, which is the real transitional form, it seems to be not earlier than the second half of the third century, in JSNab 18 (Fig. 25), in the word *šlm* at the end of line 2. In the papyri, forms 2–5 of medial *š* are attested, although form 5 is found only in the "extremely" cursive script (Yardeni 2000: 262).

The *t*, like the *š*, is a very interesting letter because neither the medial nor the final transitional forms appear before the end of the third century, in M 1 (Fig. 27), dated to AD

280. After that date, the transitional forms become the rule, except in Stiehl (Fig. 31) of AD 356 and in UJadh 309 (Fig. 48) of AD 295, which are formal texts.[27] In the latter, only the final form, in *šnt*, line 5, is transitional. Note that the "classical" and transitional forms of the letter in initial/medial position are exclusive of each other. In the papyri, where the open form of the *t* (no. 4) appears earlier than in the inscriptions, one can trace the evolution of the letter from forms 1–3 to form 4. There is indeed, in the cursive *t*, a form that is not attested in the inscriptions (see Yardeni 2000: 263).
All in all, if one was to draw the ideal alphabet of the evolved forms that appear in transitional texts, most of which, of course, already appear in the papyri, one would probably get something close to the "idealized" alphabet, which is given in Figure 19. The logic would be that the more a particular inscription contains characters that belong to this idealized alphabet, the later it is. However, the process does not exactly work like this because some letters show an early development and are used in a rather stable way throughout the corpus (the *h*, the *ḥ*, and the *y*) while others show hesitations between "classical" and evolved forms, sometimes within one single text (ʾ, *m*).

It is therefore difficult, from the inscriptions, to trace a continuous development in the use of the letter forms. Indeed, very often there are, as one would expect, mixtures of "classical" and evolved forms in the texts. The reason for this, as suggested by M.C.A. Macdonald, may be that the calligraphic version of the script used in formal inscriptions — and often successfully attempted in graffiti — co-existed with a day-to-day scribal version of the script used for documents on soft materials. He considers rightly that the "changes in the letter forms and the increasing use of ligatures seen in the formal script only make sense as the transference to stone of features developed through writing swiftly with pen and ink. There would have been no reason for them to have developed independently within the process of carving on stone".[28] The result is a "growing, but haphazard, intrusion of day-to-day scribal forms (with occasional attempts to 'monumentalize' them) into the calligraphic version of the script used for monumental inscriptions, which is imitated in graffiti".

I have tried, in this modest contribution, to gather all the examples of these "intrusions", both published and unpublished, not only to present an up-to-date assessment of the material available for the study of the development of the Nabataean script but also to help identify the intrusions, determine which are the most recurrent, and which characters offer more "resistance" and in which contexts. The material I have collected is a perfect illustration of the fact that here is no continuous development of the script, and this should make us even more careful regarding the use of palaeographic studies for dating.

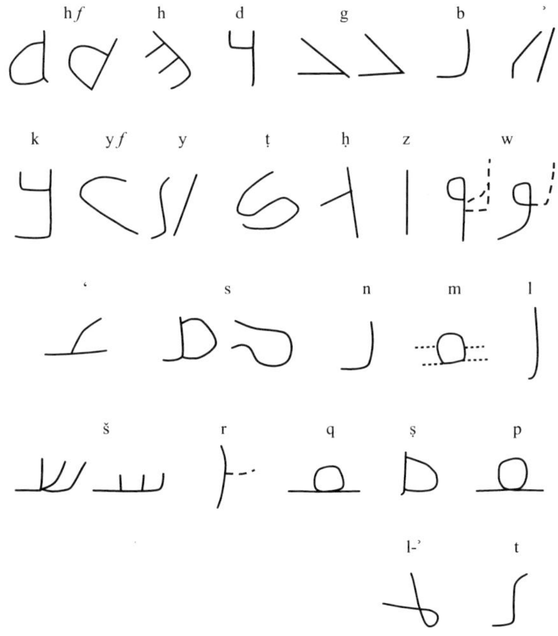

Figure 19. *The idealized forms of the evolved characters in the transitional texts.*

[27] Note that it appears in the Namārah inscription, which is also formal.

[28] See Macdonald 2003: 52–54; this volume; and forthcoming, the section entitled "The chisel and the pen"; and (this volume).

Appendix 1. Texts dated to the period AD 200–500 (third–fifth centuries), by date

Readings of the texts which appear in bold script are given in Appendix 2.

Date (AD)	*Site*	*Region*	*Reference(s)*
	Third century		
203	Taymāʾ	North-west Arabia	al-Najem & Macdonald 2009
204	ʿAvdat	Negev	*RES* 528
205–206	Wādī Mukattab	Sinai	*CIS* ii **963**
218–219	Jabal Munayjah	Sinai	*CIS* ii 2666
222–223	Wādī Ḥajjāj	Sinai	Negev 1981: no. 9
230–231	Boṣra	Ḥawrān	**B 3** (unpublished)
231–232	Wādī Fayrān	Sinai	*CIS* ii 1491
265–266	Wādī Maghārah	Sinai	NDGS 1
266–267	Wādī Maghārah	Sinai	NDGS 2
267	Madāʾin Ṣāliḥ	North-west Arabia	**JSNab 17**
267?	Madāʾin Ṣāliḥ	North-west Arabia	**JSNab 18**
Third century	Umm al-Jimāl	Southern Ḥawrān	**LPNab 41**.
275–276	Dūmat al-Jandal	North-west Arabia	**ARNA.Nab 17** (= Macdonald 2009*a*)
280	Mābiyāt	North-west Arabia	**M 1** (= al-Muraykhī, in press)
295	Umm Jadhāyidh	North-west Arabia	**UJadh 309** (unpublished)
	Fourth century		
305–306	Umm Jadhāyidh	North-west Arabia	**UJadh 297** (unpublished)
305+	Al-ʿUlā	North-west Arabia	*CIS* ii 333
306	Al-ʿUlā	North-west Arabia	JSNab 386
311–312 (or 151–152)	Umm Jadhāyidh	North-west Arabia	**UJadh 172** (unpublished)
328	al-Namārah	Ḥawrān	most recently Calvet & Robin 1997: 265–269, no. 205
356	Madāʾin Ṣāliḥ	North-west Arabia	**Stiehl** (= Stiehl 1970)
	Fifth century		
428	Sakākā	North-west Arabia	**S 1** (= al-Muraykhī, in press)
455–456	Umm Jadhāyidh	North-west Arabia	**UJadh 109** (= ThNUJ 132+133, republished in Nehmé 2009: 49–52, fig. 3).

Appendix 2. The selection of inscriptions used as examples for letter forms mentioned in this paper

The corpus of inscriptions dated to between the third and fifth centuries AD (so-called "late Nabataean"), some of which are clearly written in "transitional" characters and some less so, at present contains 116 texts. The selection of thirty-four inscriptions presented below represents therefore almost one third of the total number. They are used in this paper for examples of letter forms. In this Appendix, the readings and translations are given but commentaries are kept to a minimum. A photograph and a facsimile are provided for each text. The complete corpus will be published elsewhere, in a special volume devoted to texts from late Nabataean to early Arabic, up to AD 680. Note that fifty texts out of the 116 were examined during a workshop, which was organized in Paris in 2005 by the present author as part of a project on the development of the Nabataean script. During this workshop, a total of seventy-one texts was examined, dating from the beginning of the third century AD to the end of the reign of the Umayyad caliph Muʿāwiyah in AD 680.

Editorial sigla:

- { } enclose doubtful letters
- {.} represents an illegible letter
- [ ] enclose letters which are restored
- ---- represents a passage in which one or more letters are completely destroyed
- / between two letters indicates an alternative reading
- * indicates one of fifteen previously unpublished texts

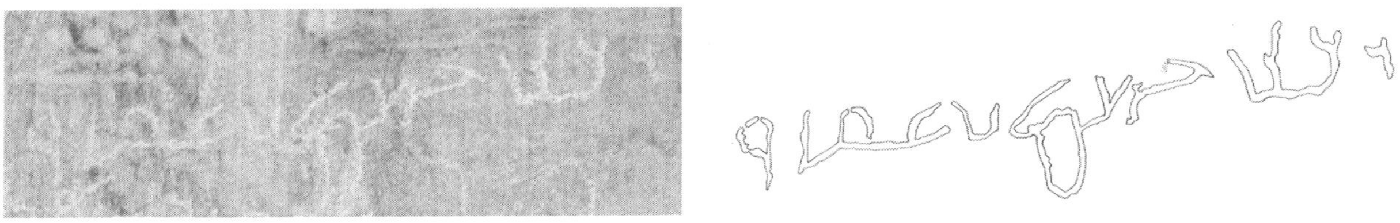

FIGURE 20. *Ar 19. (Photograph Darb al-Bakrah project, facsimile L. Nehmé).*

Ar 19 (Fig. 20)

This text comes from the site of al-ʿArniyyāt (see Fig. 7), which was visited during the Darb al-Bakrah Survey Project in 2004 during which 166 Nabataean texts were recorded. These will be published as part of the Darb al-Bakrah corpus of inscriptions.

dkyr grʿm br ʿmrw

Note the different shapes of *m*: the final *m* in *grʿm* has the "classical" Nabataean form while the medial *m* in *ʿmrw* is much more evolved and is ligatured from the preceding and to the following letters at its base. The *d* and the *k* are representative of what these letters become in the transitional texts. The *r* in *dkyr* is a simple vertical stroke.

All the letters are clear. The name *grʿm* is new in the Nabataean onomasticon whereas *ʿmrw* is common in the Nabataean inscriptions.

ARNA.Nab 17 (Fig. 21)

This text was photographed by F.V. Winnett and W.L. Reed in 1962, approximately 15 km north-north-west of Dūmat al-Jandal. It was published — but misread — by J.T. Milik and J. Starcky, from a very poor photograph, as ARNA.Nab 17, and republished in 1996 by Kh. al-Muaikil and S.A. al-Theeb who read it as two inscriptions, their nos 63 and 64. Again, no photograph was made available. The first exact reading and full commentary from an excellent photograph of the text by the Department of Antiquities and Museums of Saudi Arabia is due to M.C.A. Macdonald (2009*a*). The reading and translation given below are taken from the latter publication. The text was examined during the 2005 Paris workshop.

ʾy dkyr ʿwydw (l. 3)
br šlymw (l. 4)
khnʾ (l. 5)

FIGURE 21. *ARNA.Nab 17. (Photograph Macdonald 2009*a*: 237, pl. 1, facsimile L. Nehmé).*

dnh šnt mʾh (l. 2)
w šbʿyn (l. 1)
"Yea, may ʿwydw be remembered son of Šlymw the oracle priest. This is the year one hundred and seventy"

The text is dated to AD 275–276. Note the form of the ʾ in *ʾy*, *mʾh*, and *khnʾ*, which still has a small vertical stroke. Note also the ligature between the *m* and the *w* in *šlymw*. The medial *h* of *khnʾ* is odd because it is closed at its bottom, as a result of the continuation of the base line from the *k*. The *t* in *šnt* is also closed at its bottom but this feature is part of the letter itself, it is not the continuation of the base line.

ʿwydw is a common name in the Nabataean inscriptions. It should be noted that *šlymw* is attested elsewhere in Nabataean only in North Arabia, in ARNA. Nab 16 as reread by al-Muaikil and al-Theeb (1996: no. 35),[29] and in al-Theeb 1993: no. 21, as well as in UJadh 10, see below.

[29] Milik and Starcky's original reading of the name in ARNA.Nab.16, *šlytw*, has been corrected to *šlymw*. The examination of the original photograph taken by Winnett and Reed seems to confirm this reading (the letter is closed at the bottom), but see Macdonald 2009*a*: n. 31.

** B 3* (Fig. 22)

This text is part of the corpus of Nabataean inscriptions from Boṣra. It was found in 1956 in the ruins of the "Nabataean gate" and was kept in the Museum of the Citadel of Boṣra (where, however, it was not found by the author in 2003). It was photographed by M. Dunand in 1961 and recorded by J.T. Milik whose reading differs slightly from the one given below. The text is difficult to read because the letters are finely and lightly incised in the basalt. M.C.A. Macdonald considers it to be written in Ḥawrān Aramaic rather than in Nabataean. It seems to me, however, that the incision in the hard basalt may be responsible for the lack of some of the ligatures and most of the letters are perfectly understood and read as Nabataean.

The text was examined during the 2005 Paris workshop.

ʿbdw br {m}{.}{d/r}{d/r}{pw}
šlm št{.} mʾ{t} w
ʿšryn w ḥmš l{h}{.}{.}y
"ʿbdw son of *{m}{.}{d/r}{d/r}{pw}*, peace ! Year one hundred and twenty of the eparchy"

Figure 22. *B 3. (Photograph J. Milik's archive, facsimile L. Nehmé).*

The text is dated to AD 230–231. Note the classical form of the ʾ, the *m*, the *š*, the *t*. Of the very few letters which have transitional forms, note that final *y* in *{h}{.}{.}y*.

ʿ*bdw* is a well-known Nabataean name but the reading of his father's name is too doubtful to make any suggestion. In the word *šnt*, the *n* is missing. The letters *št* are followed by the remains of a letter, possibly an ʾ. No satisfactory explanation can be given for this sign, which may be accidental.

CIS ii 963 (Fig. 23)

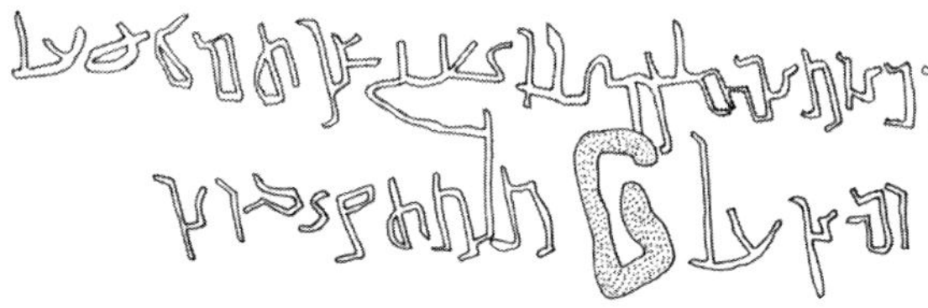

Figure 23. *A facsimile of* CIS ii *963 based on the squeeze published in* CIS *(L. Nehmé).*

This text (= *RES* 128) is known only from a squeeze published in *CIS* ii. It was examined during the 2005 Paris workshop.

dkyr tymʾlhy br yʿ{l/n}y šnt mʾh ʿl
dmyn ʿl ---- tltt qysryn

"May be remembered Tymʾlhy son of Yʿny year one hundred which equals ---- the three Caesars"

It has been suggested (*CIS* ii and Negev 1967: 253) that the signs ʿ *l* after *mʾh* were in fact the figures 5+1 but this is very unlikely. The date should therefore be understood as only "one hundred", i.e. AD 205–206, and not 100+5+1, i.e. 106 = AD 211. The three Caesars were Septimius Severus, Caracalla, and Geta, who reigned together from 198 to 211. The translation of the expression ʿ*l dmyn* as "equals" in Negev is based on his interpretation of the meaning of the verb *dmy* in Aramaic, "to resemble, be like" (Sokoloff 1990: 151/a). However, Negev does not explain how a substantive *dmyn* meaning "equals" can be derived from *dmy* and the syntax, preposition + *dmyn* + preposition, is odd. The whole phrase requires reconsideration but the date of the text is clear from what precedes it.

JSNab 17 (Fig. 24)

This well-known inscription has been widely discussed since it was first discovered by C. Huber and published by Jaussen and Savignac. I do not intend to propose a full re-examination here (for which see Healey & Smith 1989; Healey 2002; Robin 2001: 547; 2006: 324–326). It has been included in this selection because a new photograph and a new facsimile have recently been made and are given here, along with the reading and translation, for reference. This text was examined during the 2005 Paris workshop.

dnh qbrw ṣnʿh kʿbw br
ḥrtt lrqwš brt
ʿ*bdmnwtw ʾmh w hy*
hlkt py ʾl ḥgrw
šnt mʾh w štyn
w tryn byrḥ tmwz w lʿn
mry ʿlmʾ mn yšnʾ ʾl qbrw
d[ʾ] w mn yptḥh ḥšy w
wldh w lʿn mn yqbr w {y}ʿly mnh

"This is the tomb which was built by Kʿbw son of Ḥrtt for Rqwš daughter of ʿbdmnwtw his mother. And she died in al-Ḥijr in the year one hundred and sixty-two in the month of Tammūz. And may the Lord of eternity curse anyone who alters this tomb or opens it except his children and may he curse anyone who buries and removes [a body] from it."

It should be noted here that the demonstrative pronoun before *qbrw*, in line 1, is not *th*, and is therefore not

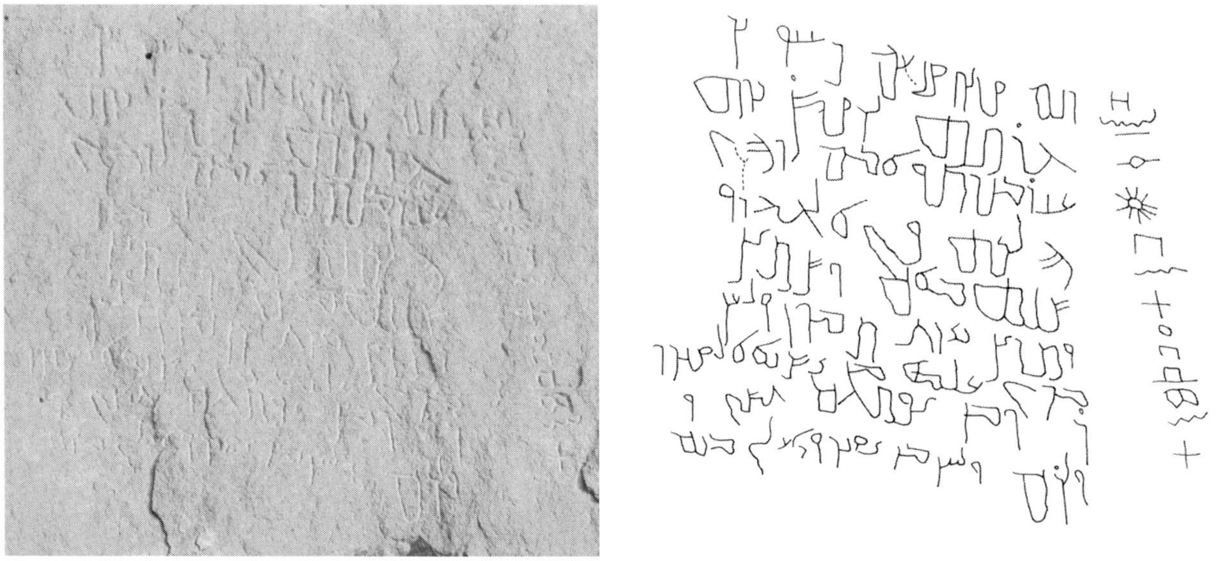

FIGURE 24. *JSNab 17. (Photograph Madāʾin Ṣāliḥ archaeological project, facsimile L. Nehmé).*

feminine (*contra* Healey & Smith 1989: 80). There is absolutely no doubt, after a minute examination of the stone, that we have here the normal masculine Nabataean demonstrative pronoun *dnh*.

The text is dated to AD 267. Note the forms of some of the letters, such as the ʾ, the *m*, both medial and final *š*, the *t*, etc., which are closer to the "classical" Nabataean form than to the transitional one. The *ḥ* and medial *h* are more evolved. Note the use of diacritical dots on the *d* (ll. 3, 8, 9).

JSNab 18 (Fig. 25)

ARNA.Nab 89–90, *RES* 1106.A, Healey 2002: 84–85.

This text was carved below and to the right of JSNab 17 and has been given much less attention by scholars. It mentions the builders of the tomb of Raqūš and is therefore either contemporary with, or a little later than, JSNab 17. It was examined during the 2005 Paris workshop.

w dkyr ʿdmn hwʾ
ktb ktbʾ dʾ bṭb w šlm
dkyr bnyʾ hnʾw w ʾḥbr{w}-
h d{y} bn{w} qbrw ʾm kʿb{w}

"And may ʿdmn who wrote this text be remembered for good and may he be secure. May the builders Hnʾw and his companions, who built the tomb of the mother of Kʿbw, be remembered"

In Healey 2002, the end of the second line is read *w bšlm* but there is no *b* in front of *šlm*. The stroke which was interpreted as the letter *b* belongs in fact to the *š* (form 3).

Note, as in JSNab 17, the "classical" form of some of the letters, the ʾ (except in *ktbʾ* and in *ʾḥbrwh*), the *w*, the *m*, the *t*, etc. The letters which are more evolved are the *h*, the *ḥ*, and the *y*. Note also the *š* in *šlm* at the end of the second line, which is clearly on its way to the evolved form of the letter.

The name *ʿdmn*, which was read *gzmn* by Jaussen and Savignac and *ʿd mn* in ARNA.Nab 90, is clear on the photograph. It is new in the Nabataean onomasticon.

LPNab 41 (Fig. 26)

CIS ii 192, *RES* 1097, Cantineau 1930–1932: 25.

This text has a Greek counterpart, which was discovered in a different place in Umm al-Jimāl. It is dated to the third century by the mention of a Jadhīmah king of Tanūkh. It was examined during the 2005 Paris workshop.

dnh npšw phrw
br šly rbw gdymt
mlk tnwḥ

"This is the nefesh of Phrw son of Šly the tutor of Gdymt the king of Tnwḥ"

Note, as in JSNab 17 and 18, the "classical" form of final *h*, of the *w*, the *m*, the *p*, the *š*, etc. Note also the peculiar form of the *t* in *tnwḥ*.

M 1 (Fig. 27)

This inscription was found in a reused position during the second season of excavations at the site of al-Mābiyāt, ancient Quraḥ, which is about 40 km south of al-ʿUlā.

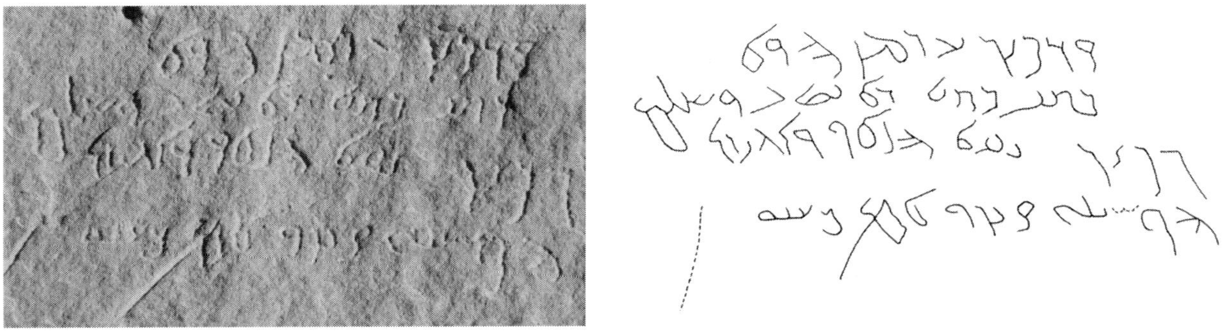

Figure 25. *JSNab 18. (Photograph Madāʾin Ṣāliḥ archaeological project, facsimile L. Nehmé).*

Figure 26. *LPNab 41. (Photograph courtesy of M.C.A. Macdonald, facsimile L. Nehmé).*

Figure 27. *M 1. (Photograph courtesy of M. al-Muraykhī, facsimile L. Nehmé).*

It has been published by M. al-Muraykhī (in press). The reading given below differs slightly from the reading proposed in the *editio princeps.*

---- šlm ʿl q[b]r r{mn}h ʾntth
brt ywsp br ʿrr dy mn qryʾ
dy mytt ywm ʿšryn w šth
bʾyr šnt mʾh w šbʿyn w ḥmš

"---- Šlm on the tomb of R{mn}h his wife daughter of Ywsp son of ʿrr, who is from Qryʾ (?), who died on day twenty-six of Iyār year one hundred and seventy-five"

The names *r{mn}h* and *ʿrr*, if the reading is correct, are new in the Nabataean onomasticon. *Ywsp* is a well-known Jewish name (Ilan 2002: 150–168), which occurs also in the recently published inscription from Taymāʾ (al-Najem & Macdonald 2009, see their commentary on the name on p. 210) as well as in UJadh 219 = ThNUJ 84 (see below) and JSNab 262. It is possible that the toponym *qryʾ* is to be equated with modern Qurayyā, in the land of Madian. The woman of whose monument this stone was part would have died in Qurayyā and was either buried or commemorated (if the stone belongs to a *nefesh*) in

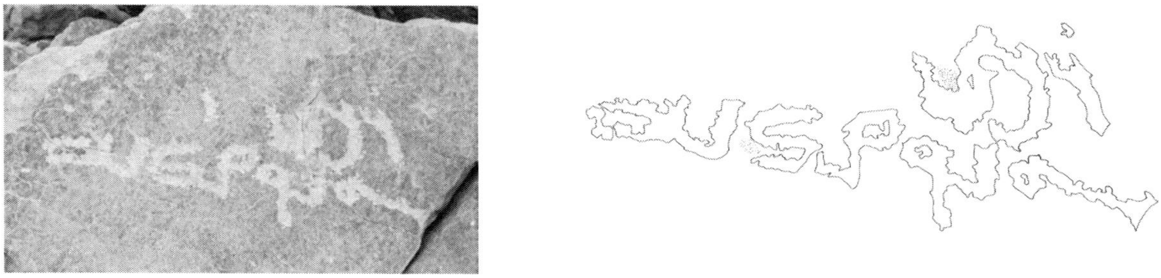

Figure 28. *QN 2. (Photograph Darb al-Bakrah project, facsimile L. Nehmé).*

Figure 29. *S 1. (Photograph courtesy of Kh. al-Muaikil, facsimile L. Nehmé).*

the region of Mābiyāt. The exact provenance of the stone is not known but if it comes from Mābiyāt itself and not from its surroundings, it means that this site was occupied before the Umayyad period.

The text is dated to AD 280.

Most of the letters in the text have an evolved form, especially the ʾ (except perhaps in *ʾntth*), final *h*, *ḥ*, medial *y* and *m*, and medial and final *t*. Note, however, the form of the *š* in *šnt*, which is on its way to a more evolved form.

*** *QN 2*** (Fig. 28)

This text comes from the site of Qāʿ an-Nuqayb (see Fig. 7), which was visited during the Darb al-Bakrah Survey Project in 2004 and at which forty-nine Nabataean texts were recorded.

dkyr ʾbwqṭ{b/n}h

Note that the possible *b* and *n* have very similar heights. Almost all the letters, except final *h*, have a "classical" Nabataean form.

The name *ʾbwqṭ{b/n}h* is new in the Nabataean onomasticon.

S 1 (Fig. 29)

This inscription, which is due to be published by Kh. al-Muaikil in *Adumatu* along with other texts from the Sakākā region, was found on an outcrop not far from Sakākā. The reading is given here with the permission of Khalīl al-Muaikil, to whom I am very grateful. This text was examined during the 2005 Paris workshop.

dkyrw mḥrbw w ʾṣḥbh
ʾl ʿšrh w ʿnymw w [w]ʾlw w ḥrtw w {ṭ/k}ḥšw
bṭbw mḥrbw br ʿwydʾlt ktb ydh ywm ʿšrh
w tmnh bʾyr šnt 2 × 100 + 100 + 20 + 3 {ʾ}{d}{.}{ḥg}-
--- ʾl ḥyrh

"May Mḥrbw and his ten companions and ʿnymw and Wʾlw and Ḥrtw and {Ṭ/K}ḥšw be remembered for good. Mḥrbw son of ʿwydʾlt wrote [with] his hand day eighteen of Iyyār the year 323*{ʾ}{d}{.}{ḥg}*---- al-Ḥīra[?]"

This text is dated to April AD 428.

For the commentary on this very interesting text, see al-Muaikil, forthcoming.

Most of the letters in this inscription have evolved forms: ʾ, *d*, final *h*, *w*, *ḥ*, *y*, *k*, *m*, *r*, *š*, *t*. The *ṣ* in *ʾṣḥbh* looks very much like a "classical" Nabataean *š* but it is different from the other *š* in the text, especially from the only initial *š*, in *šnt*. The line crossing the ligature between the *b* and the *h*, visible on the photograph, is accidental and was carved *before* the text.

The name *mḥrbw* is attested in a recently published inscription from Umm al-Jimāl, the editors of which interpreted it as the Arabic name Muḥārib (Said & al-Hamad

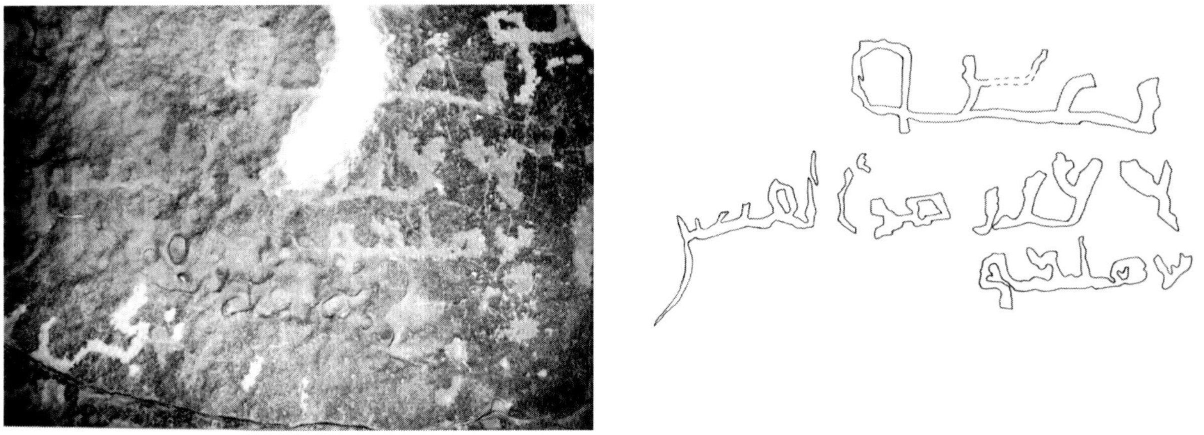

FIGURE 30. *S 3. (Photograph courtesy of Kh. al-Muaikil, facsimile L. Nehmé).*

2004: no. 2). *ʿnymw* is probably a variant of *ʿnmw*, well attested in Nabataean, as is *wʾlw*. The name *ḥrtw*, which is a variant of *ḥrt*, is attested only in one text from Umm al-Jimāl, LPNab 55, in which the reading of the name is doubtful. Neither *kḥšw* nor *ṭḥšw* are known in Nabataean and *ʿwydʾlt* is a theophoric name which occurs here for the first time, although *ʿwydw* is a well-known Nabataean name.

S 3 (Fig. 30)

This inscription was found in 1991 by S. al-Theeb on a small hill known as al-Qalʿah, about 5 km north of Sakākā, and was published by Kh. al-Muaikil (1993: no. 2, in Arabic, and 2002, no. 2, in English). Note that part of it also appears on a photograph taken in 1962 by F.V. Winnett and W.L. Reed and reproduced as plate 4 in Macdonald 2009*a*, where it was numbered ARNA.Nab 13a. This photograph was not included in Winnett and Reed 1970 and the inscription was not mentioned in the publication. This text was examined during the 2005 Paris workshop.

by{ṣ}w
dkyr mrʾlqyš
br mlkw

The *editio princeps* reads *bʿṣw* for the first name but *y* should be preferred to *ʿ*. On the second line, *dkyr* and *mrʾlqyš* should be preferred to *br ʿbd mrʾlqyš* of the *editio princeps*. It is therefore better to consider the three lines, not as a single text, but as belonging to two different inscriptions: on the one hand, the name *byṣw* which is new to the Nabataean onomasticon, and on the other, a commemorative text starting with *dkyr*. The name *mrʾlqyš* is new in the Nabataean onomasticon.

Note the peculiar form of the *ṣ* in *byṣw*, the shape of which was reconstructed from the Winnett and Reed photograph. This text shows considerably evolved characters, especially the *ʾ*, the *y*, the *m*, the *q*, the *r*, and the final *š*. The *d* and the *k* have the typical form they have in the transitional texts. This text has been dated to the fifth century by Kh. al-Muaikil (2002: 165) but this dating is not secure at all and is no more than a hypothesis. There are other texts with evolved characters on the same rock face, which can be seen in the Winnett and Reed photograph and which are discussed in Macdonald 2009*a*.

Stiehl (Fig. 31)

This inscription was found in Jedda but is said to come from Madāʾin Ṣāliḥ. It was first read by F. Altheim and R. Stiehl (1968: 305–309) and was republished by R. Stiehl in 1970. Al-Najem and M.C.A. Macdonald (2009) have recently republished it, with Stiehl's photograph and a reading which differs slightly from the earlier ones. It is this reading and translation which are given below. This text was examined during the 2005 Paris workshop.

dnh ----{š}----b{rt}ʾ dy----[30]
ʿdy---- br ḥny br šmwʾl ry{š}
ḥgrʾ ʿl mwyh ʾtth brt[31]
ʿmr{w} br ʿdywn br šmwʾl
ryš tymʾ dy mytt byrḥ
ʾb šnt mʾtyn w ḥmšyn
w ʾḥdy brt šnyn tltyn
w tmny

[30] Stiehl (1970) has *dnh [npšʾ] [w] qbrtʾ dy [ʿbd lh]* for line 1.

[31] The readings *ʿdywn* (instead of *ʿdnwn*) suggested by Al-Najem and Macdonald (2009: 213–214 and n. 35) in lines 2 and 4, and *mwyh* (instead of *mwnh*) in line 3 (first suggested by J. Starcky [1978: 47], followed by al-Najem and Macdonald [loc. cit.]) are very convincing.

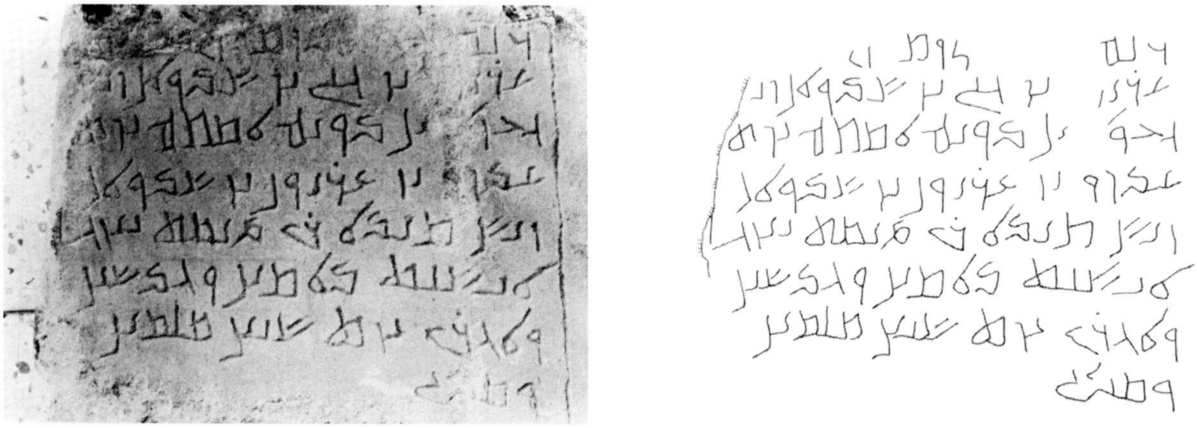

FIGURE 31. *Stiehl inscription. (Photograph R. Stiehl, facsimile L. Nehmé).*

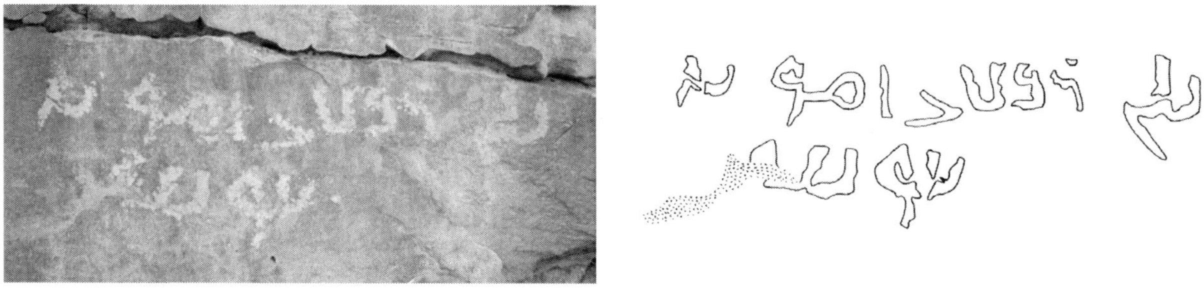

FIGURE 32. *UJadh 3. (Photograph Darb al-Bakrah project, facsimile L. Nehmé).*

"... ᶜdy[wn] son of Ḥny son of Šmwʾl chief man of Ḥgrʾ for Mwyh his wife, daughter of ᶜmrw son of ᶜdywn son of Šmwʾl chief man of Tymʾ, who died in the month of Ab in the year two hundred and fifty-one at the age of thirty-eight"

Note the "classical" form of most of the letters: ʾ, final *h*, *w*, *m*, *r*, *š*, *t*. All the examples of *d* in this text are dotted despite the fact that this was not necessary because *r* and *d* are not identical in shape. The letters which are more evolved are the *ḥ*, as well as the medial and final *y*. Note the peculiar form of the medial and final *t*, almost all the examples of which are closed at their bases. Note that none of the examples of *w* is ligatured from the preceding letter.

For the commentary on the names, see al-Najem & Macdonald 2009.

UJadh 3 = ThNUJ 48 (Fig. 32)

This text and the following ones were discovered during the Darb al-Bakrah Survey Project, in 2004, in Umm Jadhāyidh, 150 km north-west of Madāʾin Ṣāliḥ. This site contains 488 Nabataean or transitional texts, written on rock faces or boulders, among texts written in Ancient South Arabian, Arabic, Hismaic, "Thamudic", and Greek. Among the Nabataean texts, 230 had already been photographed and published by S. al-Theeb in 2002 (= ThNUJ).

There is a spring not far from the site and archaeological structures have been identified in the wadi that runs at the foot of the rocky outcrops on which the inscriptions are written. The number, variety, and sometimes very sophisticated character of the inscriptions are an indication that in antiquity this site was probably much more than a simple stop on the caravan road between Ḥegrā and Petra. Was it a sanctuary? Only a more thorough survey, or excavation, of the archaeological structures will tell.

This text was examined during the 2005 Paris workshop.

bly dkyr grmw
br ᶜwnyw

The *editio princeps* has *gzmw* instead of *grmw* but the latter is preferred here. The vertical *r* is very similar to

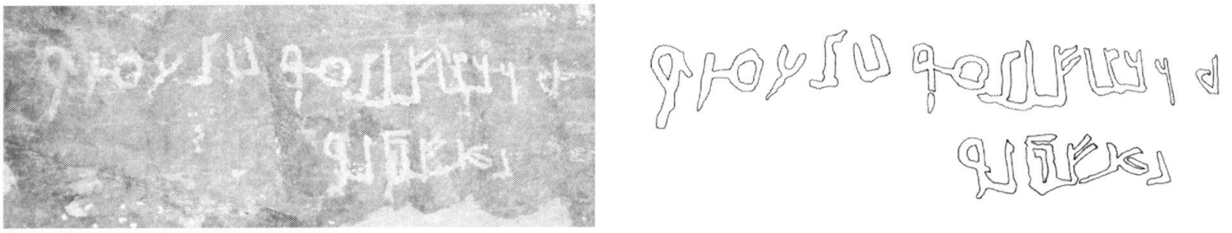

Figure 33. *UJadh 10. (Photograph Darb al-Bakrah project, facsimile L. Nehmé).*

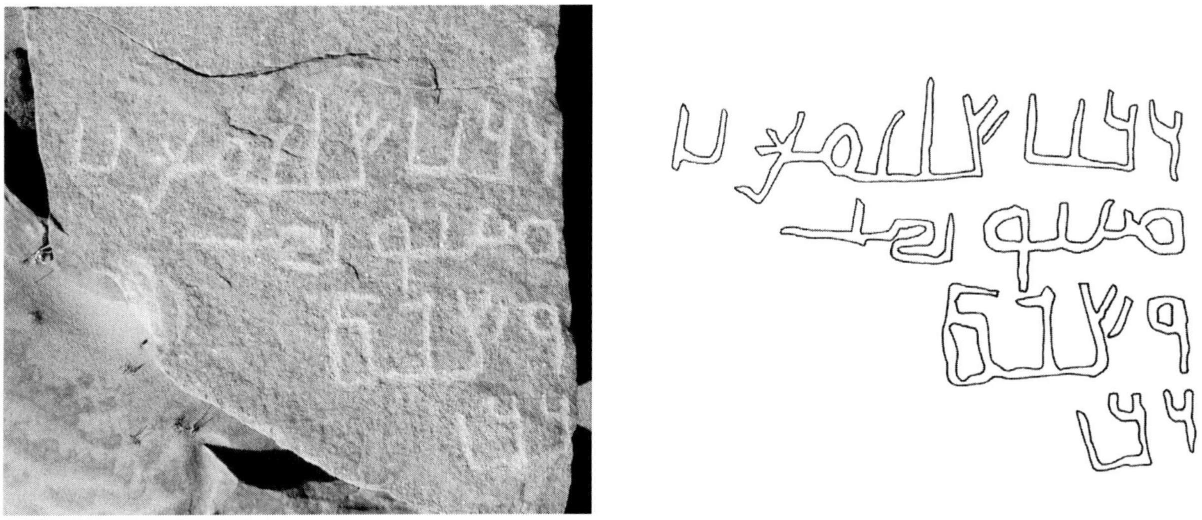

Figure 34. *UJadh 15. (Photograph Darb al-Bakrah project, facsimile L. Nehmé).*

the one in *dkyr*, except that it is not ligatured from the *g* on the right.

Note the forms of the medial and final *y*, the triangular form of the *g*, and the form of the *r*, which is a simple vertical stroke. The *d* is dotted.

The name *grmw* is very well known in Nabataean. The name *ʿwnyw* is new in the Nabataean onomasticon but may be compared to *ʿwnw*, attested in two inscriptions, one from al-ʿUlā, JSNab 202, in which the name could also be read *ʿwyw*, and one from Ḥegrā, JSNab 285, the reading of which was checked on the original, *ʿwnw* being the best one.

UJadh 10 = ThNUJ 38 (Fig. 33)

This text was examined during the 2005 Paris workshop.

dkyr šlymw br yʿmrw
bṭb šmnw

This text and the next share the same characteristics. However, they cannot have been written by the same person because their authors do not have the same patronym and the name in UJadh 15 is better read as *Šlym{n}*.

Note the occurrence in this text of both the "classical" and evolved forms of *m*. The *š* has a "classical" form and the script has a very upright aspect.

The name *šlymw* is attested elsewhere in Nabataean only in ARNA.Nab 17 and other texts from North Arabia (on which see the commentary under ARNA.Nab 17). The name *yʿmrw* occurs only once elsewhere, in *CIS* ii 195, from Umm al-Raṣāṣ, but it is attested in the form *yʿmr* in an inscription from Jabal Ṣarbūṭ Thulaythah, south-west of Tabūk (al-Theeb 1993: no. 49). The name *šmnw* is new in the Nabataean onomasticon.

UJadh 15 = ThNUJ 30 (Fig. 34)

This text was examined during the 2005 Paris workshop.

dkyr šlym{n} br
mʿnw bṭb
w š{l}m
dky

The sign above *dkyr* was read as *[bl]{y}* in the *editio princeps* but the letter visible on the photograph can hardly be a final *y*. The last letter of the first name is

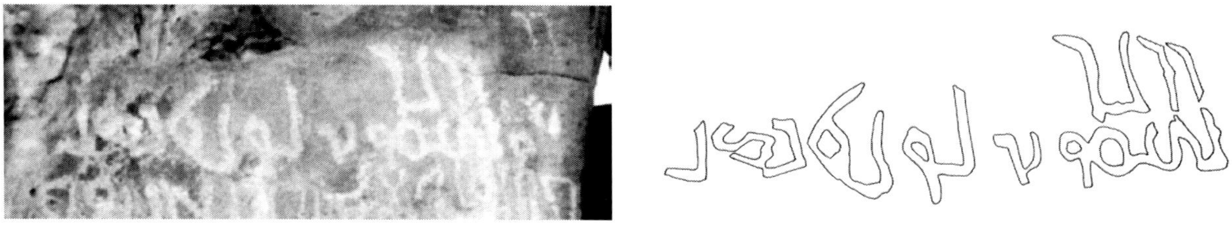

Figure 35. *UJadh 19. (Photograph al-Theeb 2002: 289, facsimile L. Nehmé).*

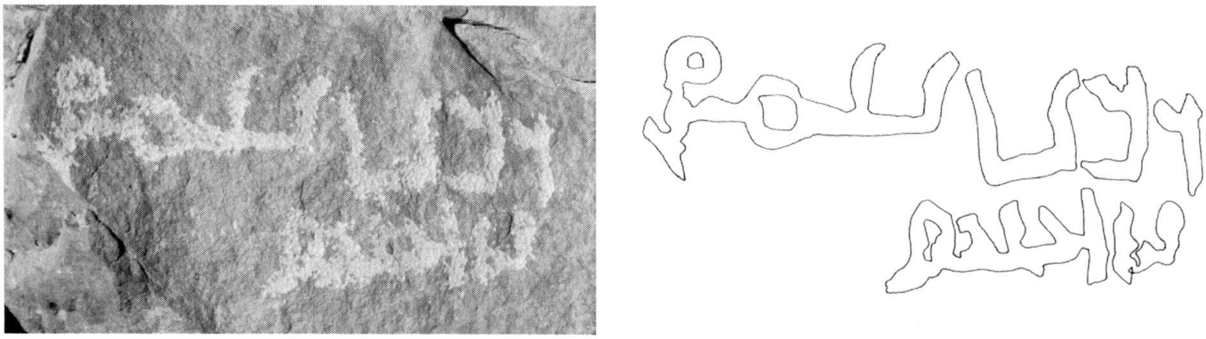

Figure 36. *UJadh 90. (Photograph Darb al-Bakrah project, facsimile L. Nehmé).*

doubtful because it is a peculiar *n*, but it cannot really be read as a *w* because there is no loop and because there is a right angle at the base. Other letters have a peculiar form: the *ṭ*, the *l* hooked at the top, and the final *m* of *šlm*. Note the very vertical character of the *y*. The *m*, except the final one in *š{l}m*, is very evolved. On the fourth line, there is probably an unfinished *dkyr*.

The name *šlym{n}* is new in the Nabataean onomasticon whereas the name *mʿnw* is very well attested in it.

UJadh 19* = *ThNUJ 34 (Fig. 35)

This text was examined during the 2005 Paris workshop.

dkyr
{l/r}ḥymw br lwyʾ bṭb

The *k* of *dkyr* does not have a horizontal line at its base. The first letter of the second line is more like a *l* because it is joined to the following letter (this would not have been the case if it was a *r*) and it has no bar at the top, unlike the other examples of *r*. The name is therefore either *lḥymw* or *rḥymw*, both of which are new in the Nabataean onomasticon.[32] As suggested by M.C.A. Macdonald, the name *lḥm* occurs in Safaitic and laḥīm "plump" or luḥaym "little plump one" are conceivable names.

Note the form of the *ḥ*, which is between the "classical" and evolved forms of the letter. Only the *y* and the *m* are clearly evolved.

The names *{l/r}ḥymw* and *lwyʾ* have not been found before in Nabataean. The latter may be a Jewish name (see Ilan 2002: 182–185).

* ***UJadh 90*** (Fig. 36)

dkyr ynmw
br ḥ{b/n}y{b/n}w

The *d* and the *k* are typical of the transitional texts and most of the characters have evolved forms, particularly the last *w*, which is very close to Arabic.

The patronym is either *ḥnynw* or *ḥbybw*, both well attested in Nabataean. The name *ynmw* has been found before only in JSNab 285, mentioned above, which was checked on the original.

UJadh 105* = *ThNUJ 128 (Fig. 37)

This text was examined during the 2005 Paris workshop.

dkyr šʿdw
br ʿbdʾyš
bšlm

All examples of *d* in this text are dotted. Note the evolved forms of the ʾ, the *y*, and the *š*. The final *m* in *šlm* has retained its classical form.

---

32 For *rḥymw*, see *rḥym*, in al-Theeb 1993: no. 60; *rḥmh* in JSNab 304, the reading of which was checked on the original; *rḥmy* in JSNab 355; finally *rḥymbl* in *RES* 1427D from Petra (this name was listed under *rḥymbʿl* by A. Negev [1991: no. 1066] but it is definitely *rḥymbl*). This inscription was photographed by the author in 2003 and the reading was checked.

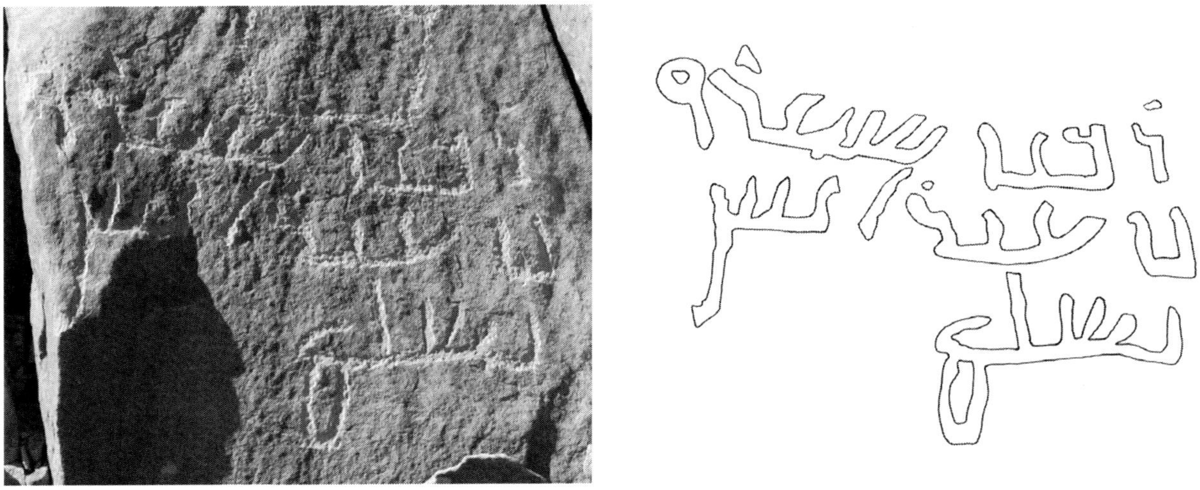

Figure 37. *UJadh 105. (Photograph Darb al-Bakrah project, facsimile L. Nehmé).*

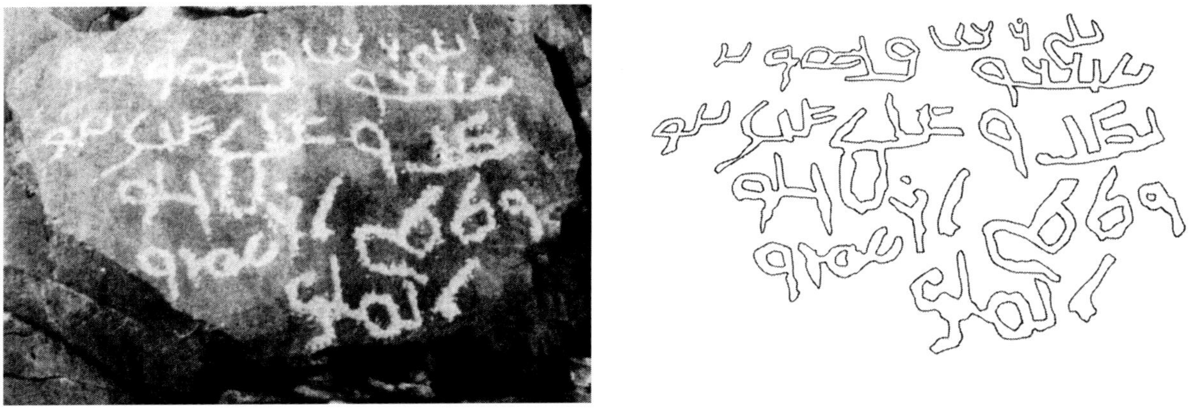

Figure 38. *UJadh 109. (Photograph al-Theeb 2002: 311, facsimile L. Nehmé).*

The name *šʿdw* is very well attested in Nabataean. As for *ʿbdʾyš*, it may be considered as a defective form of *ʿbdʾysy*, a theophoric name meaning "the servant of Isis", which is well attested in Petra (with a *samekh*), in *RES* 1431B as reread by J.T. Milik and by the present author on the original, in *RES* 1382 and 1435 (checked by the author on the original), in Milik & Starcky 1975: 128–129, pl. 47/2 (also checked on the original), and finally, perhaps, in *CIS* ii 481, as reread by J.T. Milik, an inscription for which, however, no photograph is available.

UJadh 109 = ThNUJ 132–133 (Fig. 38)

This text was examined during the 2005 Paris workshop. It was republished as part of an article on the Roman period in north-west Arabia in Nehmé 2009: 49–52, fig. 3.

bly dkyr phmw br
ʿbydw šlm šnt 2 × 100
+ 100 + 20 + 20 + 10 ʾdḥlw
ʿmrw
ʾlmlk

"Yea, may Phmw son of ʿbydw be remembered [and] may he be secure, year 350 [when] they introduced ʿmrw [ʿAmrū] the king"

This text is dated to AD 455–456. See the commentary in Nehmé 2009.

Two of the three examples of *d* are clearly dotted and it is just possible that the *d* of *ʿbydw* has a very small and faint dot above it. Note the evolved form of the *ʾ*, the *h*, the *w* when ligatured from the right, the *ḥ*, the *y*, the *m*, and the *r*. However, the *p*, the *š*, and the

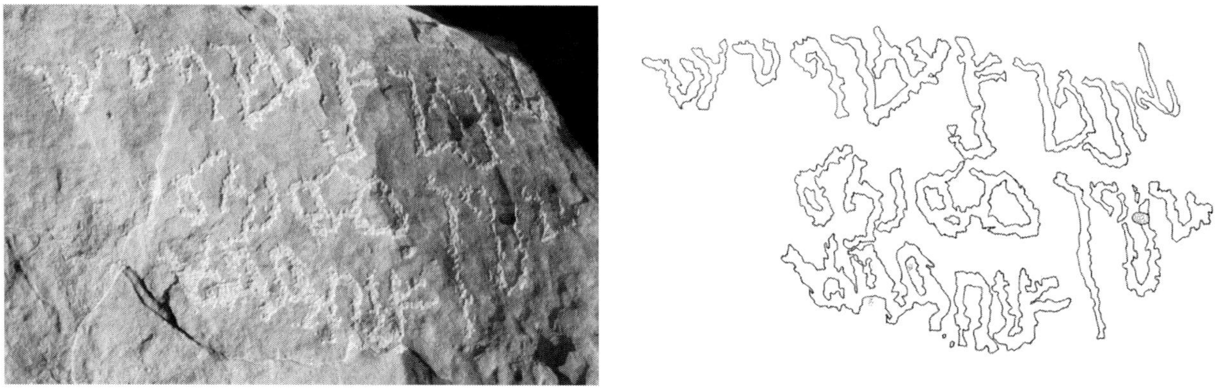

Figure 39. *UJadh 172. (Photograph Darb al-Bakrah project, facsimile L. Nehmé).*

final *m* in *šlm*, are "classical" Nabataean.

The final *k* in *ʾlmlk* is different from any of the shapes which are shown in the tables. It should be noted that it has kept a more or less horizontal line at its bottom, which normally disappears in the classical Nabataean final *k* but is present in the Kufic form. Note the use of the Arabic definite article in *ʾlmlk*.

ʿbydw is very well attested in Nabataean. The name *phmw* recurs only in UJadh 375 (see below) and in two unpublished inscriptions from al-ʿUdhayb, north of al-ʿUlā.

ʾdḥlw is of course the third person plural perfect of the *ʾafʿal* form of the Arabic verb *dakhala*, "to cause to enter, to introduce".

* ***UJadh 172*** (Fig. 39)

bly dkyr šʿydw br ʿbd-
ʿdnwn mʾbyʾ
šnt 100 + 100 + 5 + 1 + 1 (or, less probably, *20 + 20 + 5 + 1 + 1*)[33]

There is doubt about the date because the first two figures could be read either as two 20s or as two 100s. However, if each of these signs were to be read as 20, one would expect the loop to have remained opened on the left (for examples, see Milik & Seyrig 1958: fig. 2). If the signs are 20s, the date is year 47 of the eparchy, i.e. AD 152–153. If the signs are 100s, the date is year 207 of the eparchy, i.e. AD 312–313. A date in the middle of the second century would fit more with the script of the text, in which none of the letters is evolved. The characters are "classical" Nabataean. Finding them at the beginning of the fourth century is surprising, especially if we compare this text with the script of UJadh 297, which comes from the same site (see below and Fig. 45), dated to AD 305–306.

Šʿydw is a common name in the Nabataean inscriptions and *ʿbdʿdnwn* is attested in JSNab 38, carved on the monumental tomb IGN 100 in Ḥegrā.

Mʾbyʾ is either a *nisba* form, like *ḥgryʾ* in JSNab 150, or a professional name. The most obvious explanation is that it is the *nisba* form derived from the toponym *mʾb*, thus perhaps, as suggested by M.C.A. Macdonald, "the Moabite". In Hebrew, it is spelt with a *w* (*môʾabî*), and *mwbyʾ* in JSNab 157 is supposed to be the Nabataean form, but one can envisage the shortening of the initial vowel to produce *mʾbyʾ*.

* ***UJadh 178*** (Fig. 40)

dkyr ʿnmw br zk{yw}
bṭb w šlm

The last two letters of the patronym may safely be read as *zk{yw}* because the same name and patronym appear in another text of Umm Jadhāyidh, probably written by the same man. Note the evolved forms of the *y*, the medial *m*, and the *š*. The *d* and the *k* have the forms they normally have in the transitional texts. The only *d* in the text is dotted although it does not have the same shape as the *r*.

ʿnmw is a common name in the Nabataean inscriptions.

UJadh 219 = ThNUJ 84 (Fig. 41)

This text was examined during the 2005 Paris workshop.

dkyr ywsp
br ʿnmw
bṭb w šlm

The letters in this text are not as ligatured as they normally would be (between the *n* and the *m* and between the *b* and the *ṭ*). It seems that the *k* had a dot over it but it is also possible that it is a crack in the stone. Note the form of

[33] M.C.A. Macdonald has suggested to me that there are two strokes on top of the five, not just one, as I initially thought.

Figure 40. *UJadh 178. (Photograph Darb al-Bakrah project, facsimile L. Nehmé).*

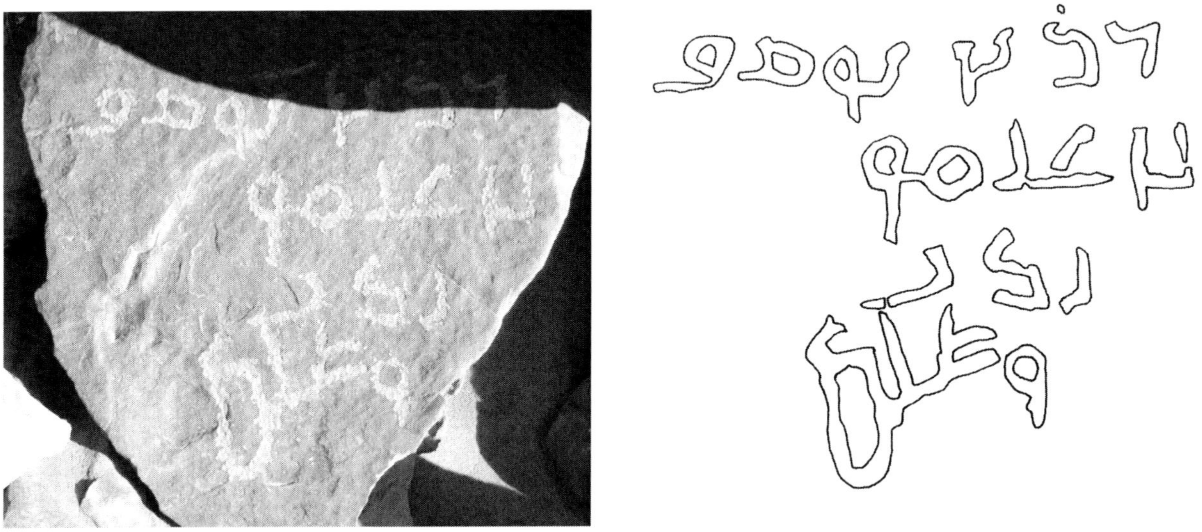

Figure 41. *UJadh 219. (Photograph Darb al-Bakrah project, facsimile L. Nehmé).*

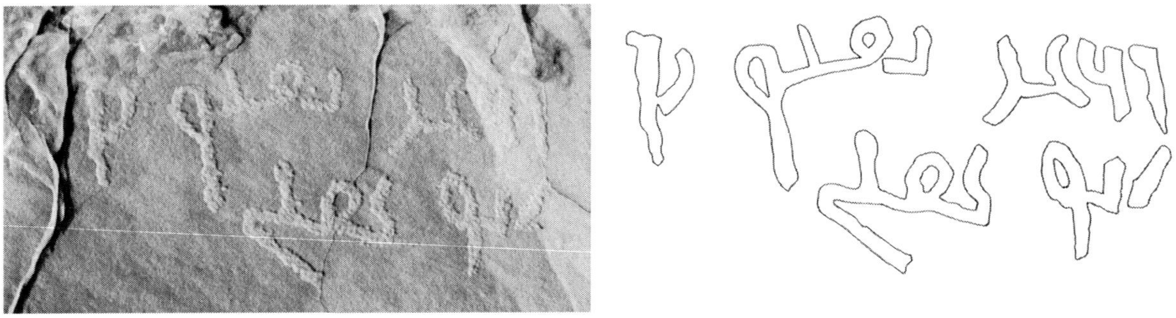

Figure 42. *UJadh 222. (Photograph Darb al-Bakrah project, facsimile L. Nehmé).*

the *s*, which is different from that in classical Nabataean. Note finally the evolved forms of *w*, *y*, and medial *m*. The ʿ clearly sits on the line.

On *ywsp*, see M 1 above. On *ʿnmw*, see the previous text.

**** UJadh 222*** (Fig. 42)

dkyr {b/n}pnw br
ʾbw ypny

The medial *k* in *dkyr* does not have a horizontal bottom line and looks therefore like a transitional *d*. Note the

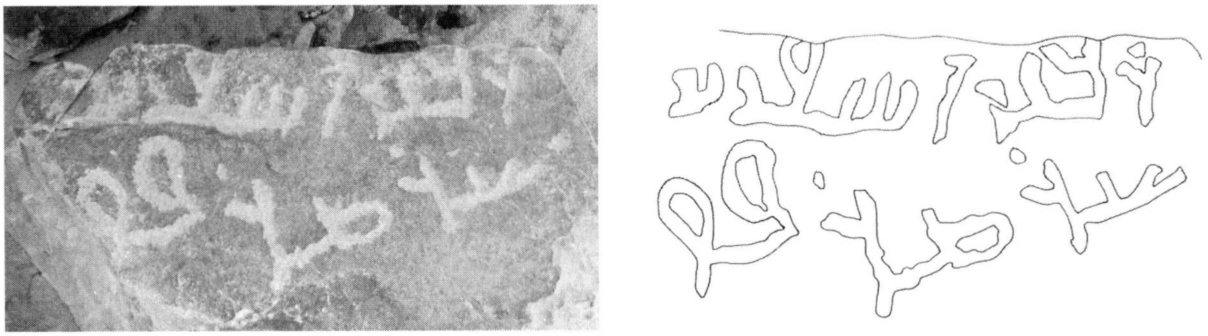

Figure 43. *UJadh 248. (Photograph Darb al-Bakrah project, facsimile L. Nehmé).*

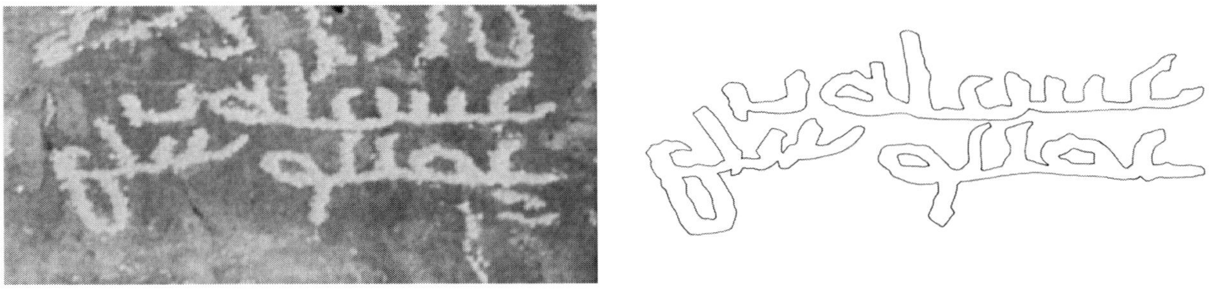

Figure 44. *UJadh 266. (Photograph Darb al-Bakrah project, facsimile L. Nehmé).*

form of the *r* in *dkyr*, which is almost exactly that of early Arabic. Note also the form of the medial *p*, which is very close to Arabic as well as the evolved forms of ʾ and *y*.

Neither of the names is attested in Nabataean. Names formed with *ʾbw* are rare in the Nabataean onomasticon.

*** *UJadh 248*** (Fig. 43)

dkyr {ʾ}š{p}r br
ʿbdṣdpw

Note the evolved form of *ṣ* as well as of the ʾ, the *y*, and especially the *š*. All three examples of *d* have dots above them.

The names, if correctly read, are new to the Nabataean onomasticon. They may be derived from the roots Š-F-R and Ṣ-D-F, both of which exist in Arabic.

*** *UJadh 266*** (Fig. 44)

ʿšylh br
ʿmyyw šlm

Note the very evolved form of the *š*, especially the first one, as well as the form of the final *h*, the *y*, and the medial *m*. The final *m* in *šlm* has kept its classical Nabataean form.

The name *ʿšylh* is new in the Nabataean onomasticon and may be derived from the Arabic root ʿ-S-L, *ʿusaylah* meaning, among other things, a drop of honey. This interpretation of the name would mean that final *-h* here represents a *tāʾ marbūṭah* and this would possibly be the earliest example of its use, the other earliest example being the Jabal Says graffito (see the contribution of P. Larcher, this volume). According to A. Negev (1991: no. 903), the name *ʿmyyw* is attested 289 times in Sinai.[34]

*** *UJadh 297*** (Fig. 45)

This text was examined during the 2005 Paris workshop.

ʿwpw br
wʾ{y}lw ktb ydh
šnt 2x100

"ʿwpw son of Wʾylw wrote with his hand, year 200"

The date is written in the form of two units attached to the symbol of the hundred, thus 200, and the text is dated to AD 305–306. It is very clear and there is no ambiguity in the reading.

[34] Note that, *contra* Negev, the name is not attested in ARNA.Nab 20, where it is *ʾmyw*.

Figure 45. *UJadh 297. (Photograph Darb al-Bakrah project, facsimile L. Nehmé).*

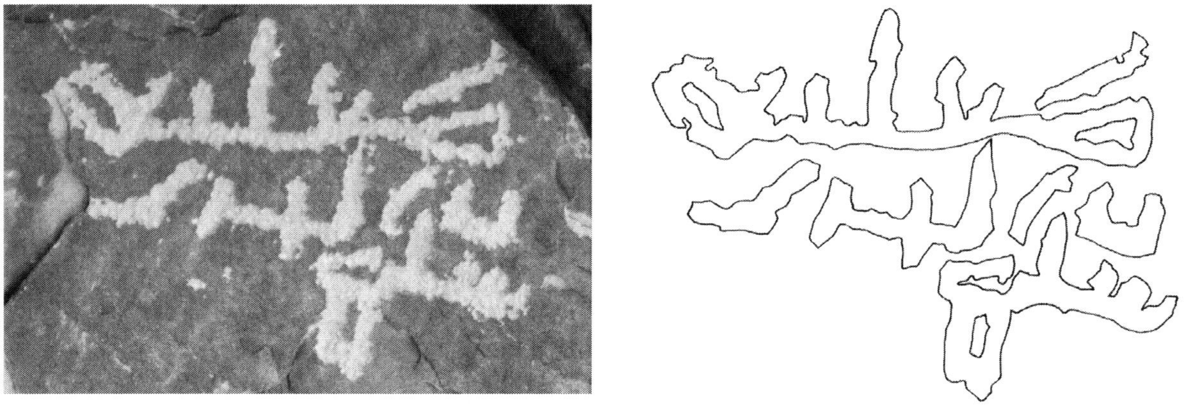

Figure 46. *UJadh 298. (Photograph Darb al-Bakrah project, facsimile L. Nehmé).*

Most of the letters of this text are very evolved: the ʾ, the final *h*, the *y*, and the medial and final *t*.

The name *ʿwpw* is new in the Nabataean onomasticon, but ʿAwf is common in Arabic. The name *wʾylw* has not been found before, though *wʾylt* occurs once and *wʾlw* is very common. It is very probable that it represents a diminutive form *wuʾayl.

*** *UJadh 298*** (Fig. 46)

This text was examined during the 2005 Paris workshop.

ṯʿlbh
br ʾlḥrt
šlm

Note the very evolved forms of the ʾ, the final *h*, the *ḥ*, the *ṯ*, the *r*, the *š*, and the final *t*. The final *m* has retained its classical form.

The name *ṯʿlbh* is new in the Nabataean onomasticon and its origin is difficult to trace. The name *ʾlḥrt* occurs twice in JSNab 382 and is very well known in early Arabic.

*** *UJadh 299*** (Fig. 47)

This text was examined during the 2005 Paris workshop.

ʾbw ʿmrw br ʿšylh
šlm šlm

Note the very evolved forms of the ʾ, the final *h*, the *y*, the *m,* the *r*, and the *š*. The two examples of final *m* in *šlm* have retained their classical form.

ʿmrw is a common name in the Nabataean inscriptions and it is preceded here, for the first time, by *ʾbw*. On *ʿšylh*, see above, UJadh 266.

*** *UJadh 309*** (Fig. 48)

bly dkyr šly br ʾwšw
br ʾlḥ{b/n}h bṭb w šlm
w ktbʾ dnh ktb
ywm
ḥd btšry šnt
mʾt w tšʿyn

"Yea! May Šly son of ʾwšw son of ʾlḥnʾ be remembered for good and may he be secure. And this writing he wrote the first day of Tišrī, year one hundred and ninety"

The text, which is a commemorative text of the *dkyr* + *šlm* type, is dated to the month of Tišrī year 190 of the Province, i.e. AD 295.

Part of the first letter of UJadh 309 is carved over part of the last letter of UJadh 317, a Nabataean text written

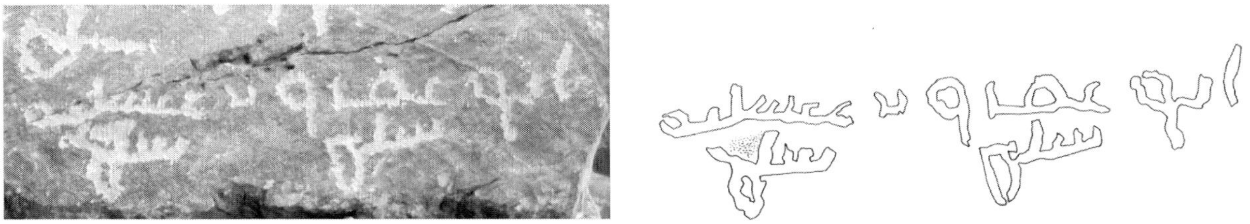

FIGURE 47. *UJadh 299. (Photograph Darb al-Bakrah project, facsimile L. Nehmé).*

FIGURE 48. *UJadh 309. (Photograph Darb al-Bakrah project, facsimile L. Nehmé).*

FIGURE 49. *UJadh 317 and UJadh 309. (Photograph Darb al-Bakrah project).*

in more "classical" characters, which reads *zbdʿdnwn br ʾnʿm* (Fig. 49). The size of the text is not known.

The reading is perfectly clear and there is no ambiguity apart from the possible confusion between *b* and *n* in the name *ʾlḥ{b/n}h*. This name does not occur in other Nabataean texts, although *ḥnh* is attested once, in JSNab 31, on tomb IGN 64 in Madāʾin Ṣāliḥ, possibly as a woman's name, Ḥannah. The Arabic roots from

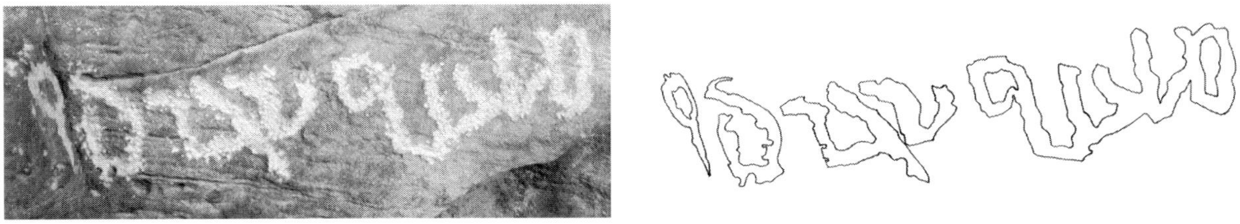

Figure 50. *UJadh 360. (Photograph Darb al-Bakrah project, facsimile L. Nehmé).*

Figure 51. *UJadh 367. (Photograph Darb al-Bakrah project, facsimile L. Nehmé).*

which the name could be derived are either Ḥ-N-N, "to be affected by an intense emotion", or Ḥ-B-B, "to be in love", or L-Ḥ-N, "to be intelligent". The alternatives, Ḫ-B-B, "to amble, to trot", Ḫ-N-N, "to speak nasally", and L-Ḫ-N, "stench", are less appropriate roots for a name. The names Ḥunn, Ḥabbah, and Ḥabbā (but not *Al-Ḥabbah), can be found in W. Caskel's *Ǧamharat an-Nasab* (1966), but it does not give any name derived from the root L-Ḥ-N. Theoretically, the letters *ʾ-l* at the beginning of the name could represent either the article or the *ʾafʿal* form of a root beginning with *l*, but the latter hypothesis is less likely despite the fact that the name *ʾlḥn*, "more intelligent" (?), is attested in Safaitic in WH 1322 and 1328 while *ʾbʾlḥn* is attested in Dadanitic in JSLih 291. The final *h* in the name is probably a *tāʾ marbūṭah*. *ʾlḥnh* is the name of the grandfather of the author. His son and grandson, *ʾwšw* and *šly*, bear names which are common in the Nabataean inscriptions.

Note that *šnt* is written with an open form of the final *t* (no. 4 on Fig. 4) whereas the final *t* in *mʾt* (a word which is normally spelt *mʾh* in Nabataean in the absolute state) is written with a looped *t*.

Note the forms of the *ʾ*, the final *h*, the *ḥ*, the medial and final *y*, the *m*, and the *š*. The *d* and the *k* have the typical form of these letters in the transitional script. The medial *t* in *ktbʾ* and *ktb* has a classical Nabataean form as has the final *m*.

UJadh 360 = ThNUJ 62 (Fig. 50)

This text was examined during the 2005 Paris workshop.

mʿnw br grmw

Mʿnw and *grmw* are well-known Nabataean names.

*** *UJadh 367*** (Fig. 51)

This text was examined during the 2005 Paris workshop.

ḥbšh
br ʿbdʾlʾšḥn
br ʿbdʾlʾšḥn

The father and grandfather of the author have the same name. This text offers two very nice examples of *lām-alif* combined into one grapheme. The letters *ʾ*, *ḥ*, *r*, and *š* have very evolved forms.

The name *ḥbšh* is new in the Nabataean onomasticon. It may be derived from the Arabic root Ḥ-B-Š, the basic meaning of which is "to collect". The name of the father and of the grandfather is also new. The name consists of the Arabic article and the Arabic root Š-Ḥ-N, which has several meanings in the *ʾafʿal* form. The fact that it is preceded by *ʿbd* suggests that *ʾšḥn* is a divine name but I have found no parallel for this.[35]

[35] The closest is the name *šhr*, which is an epithet of the moon god (Cross 1986: 391), but the final *n* is clear.

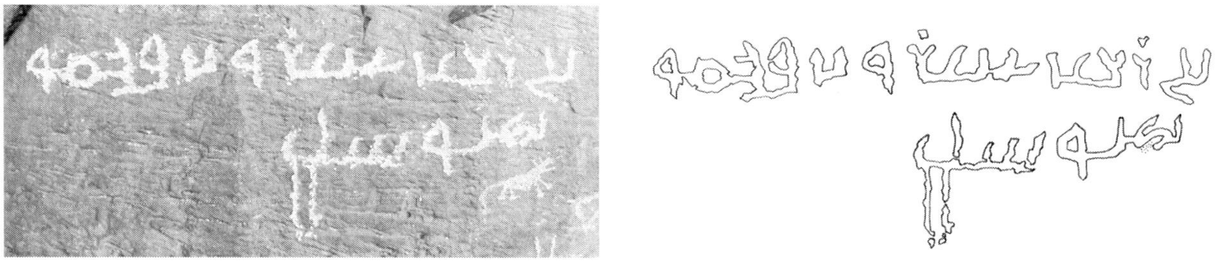

Figure 52. *UJadh 375. (Photograph Darb al-Bakrah project, facsimile L. Nehmé).*

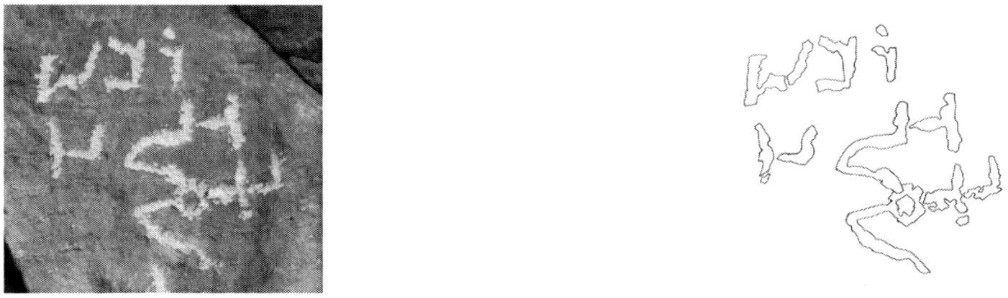

Figure 53. *UJadh 405. (Photograph Darb al-Bakrah project, facsimile L. Nehmé).*

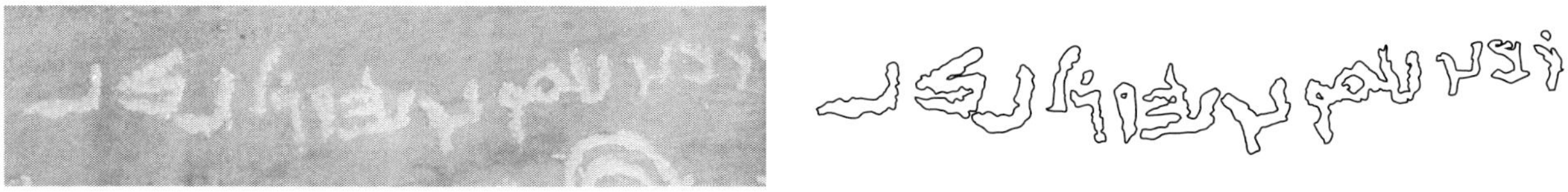

Figure 54. *ʿUlā 1. (Photograph Madāʾin Ṣāliḥ archaeological project, facsimile L. Nehmé).*

UJadh 375 = ThNUJT 38 (Fig. 52)

This text was examined during the 2005 Paris workshop.

bly dkyr ʿbydw br phmw

bṭb w šlm

Almost all the letters of this text are evolved, except the final *m* in *šlm*. The two examples of *d* are dotted.

ʿbydw is a common name in the Nabataean inscriptions. On *phmw*, see UJadh 109 above.

UJadh 405 = ThNUJ 145 (Fig. 53)

dkyr

ḥ{b/n}y br

nḥmy

The *d* is dotted. The letters of this text all have an evolved form.

The name *ḥny* is attested only once in Nabataean, in the Stiehl inscription, but it is common in Safaitic. A name *ḥby* was thought to be attested in JSNab 14 but it was reread as *ḥpy* by J. Healey (1993: 148). *Nḥmy* may be the Hebrew name *nḥmyʾ/nḥmyh* (Ilan 2002: 197). The alternative reading *bḥmy* is less satisfactory and the second letter is not likely to be a *g*, which allows us also to exclude *ngmy*.

* **ʿUlā 1** (Fig. 54)

dkyr {l/n}ḥmw br yhwdʾ bṭb

The two examples of *d* are dotted. The letters in this text are almost all evolved.

The name *lḥmw* (see UJadh 19 for *lḥm*) is attested in JSNab 136 from Madāʾin Ṣāliḥ but if we consider that the first letter of the name in ʿUlā 1 is rather short for a *l* and in view of the Jewish patronym, the Jewish name *nḥmw* is more likely (see *nḥmy* in the previous text). *Yhwdʾ* is of course the Hebrew name Judah (Ilan 2002: 112–125), which is attested in the form *yhwdh* in the signatures of the Starcky Papyrus, line 39 (Yardeni 2001: 129, 133).

Appendix 3. Names contained in the inscriptions listed in Appendix 2

Previously unpublished inscriptions are followed by *.

ʾbw ypny	UJadh 222*
ʾbw ʿmrw	UJadh 299*
ʾbw qṭ{b/n}h	QN 2*
ʾwšw	UJadh 309*
ʾlḥ{b/n}h	UJadh 309*
ʾlḥrt	UJadh 298*
{ʾ}š{p}r	UJadh 248*
by{ṣ}w	S 3
{b/n}pnw	UJadh 222* (see also under *{n/b}pnw*)
gdymt	LPNab 41
grmw	UJadh 3 (= ThNUJ 48), UJadh 360 (= ThNUJ 62)
grʿm	Ar 19*
hnʾw	JSNab 18
wʾ{y}lw	UJadh 297*
[w]ʾlw	S 1
zk{yw}	UJadh 178*
ḥ{b/n}y	UJadh 405 (= ThNUJ 145)
ḥ{b/n}y{b/n}w	UJadh 90*
ḥbšh	UJadh 367*
ḥny	Stiehl
ḥrtw	S 1
ḥrtt	JSNab 17
{ṭ/k}ḥšw	S 1 (see also under *{k/ṭ}ḥšw*)
ṭʿlbh	UJadh 298*
yhwdʾ	ʿUlā 1*
ywsp	M 1, UJadh 219 (= ThNUJ 84)
ynmw	UJadh 90*
yʿmrw	UJadh 10 (= ThNUJ 38)
yʿny	*CIS* ii 963
{k/ṭ}ḥšw	S 1 (see also under *{ṭ/k}ḥšw*)
kʿbw	JSNab 17, 18
lwyʾ	UJadh 19 (= ThNUJ 34)
{l/n}ḥmw	ʿUlā 1* (see also under *{n/l}ḥmw*)
{l/r}ḥymw	UJadh 19 (= ThNUJ 34) (see also under *{r/l}ḥymw*
mwyh	Stiehl
mḥrbw	S 1
mlkw	S 3
mʿnw	UJadh 15 (= ThNUJ 30), UJadh 360 = (ThNUJ 62)
mrʾlqyš	S 3
nḥmy	UJadh 405 (= ThNUJ 145)
{n/b}pnw	UJadh 222* (see also under *{b/n}pnw*)
{n/l}ḥmw	ʿUlā 1* (see also under *{l/n}ḥmw*)
ʿbdʾyš	UJadh 105 (= ThNUJ 128)
ʿbdʾlʾšḥn	UJadh 367*
ʿbdw	B 3*
ʿbdmnwtw	JSNab 17
ʿbdʿdnwn	UJadh 172*
ʿbdṣdpw	UJadh 248*
ʿbydw	UJadh 109 (= ThNUJ 132–133), UJadh 375 (= ThNUJT 38)
ʿdywn	Stiehl
ʿdmn	JSNab 18
ʿwydʾlt	S 1
ʿwydw	ARNA.Nab 17
ʿwnyw	UJadh 3 (= ThNUJ 48)
ʿwpw	UJadh 297*
ʿmyyw	UJadh 266*
ʿmrw	Stiehl, Ar 19*, UJadh 109 (= ThNUJ 132–133)
ʿnymw	S 1
ʿnmw	UJadh 178*, UJadh 219 (= ThNUJ 84)
ʿrr	M 1
ʿšylh	UJadh 266*, UJadh 299*
phmw	UJadh 109 (= ThNUJ 132–133), UJadh 375 (= ThNUJT 38)
phrw	LPNab 41
{r/l}ḥymw	UJadh 19 (= ThNUJ 34) (see also under *{l/r}ḥymw*)
r{mn}h	M 1
rqwš	JSNab 17
šly	LPNab 41, UJadh 309*
šlymw	ARNA.Nab 17, UJadh 10 (= ThNUJ 38)
šlym{n}	UJadh 15 (= ThNUJ 30)
šmwʾl	Stiehl
šmnw	UJadh 10 (= ThNUJ 38)
šʿdw	UJadh 105 (= ThNUJ 128)
šʿydw	UJadh 172*
tymʾlhy	*CIS* ii 963

Sigla and abbreviations

Ar	Site name: al-ᶜArniyyāt (see Fig. 7).
ARNA.Nab	Nabataean inscriptions published or mentioned in Winnett & Reed 1970.
B	Site name: Boṣra (see Fig. 7).
CIS ii	*Corpus Inscriptionum Semiticarum.* Pars II. *Inscriptiones Aramaicas continens.* Paris: Imprimerie nationale, 1889–1954.
CIS v	*Corpus Inscriptionum Semiticarum.* Pars V. *Inscriptiones Saracenicas continens.* Tomus 1. *Inscriptiones Safaiticae.* Paris: Imprimerie nationale, 1950–1951.
Ir	Site name: ᶜĪrīn.
JSLih	"Liḥyanite" (Dadanitic) inscriptions published in Jaussen & Savignac 1909–1914.
JSNab	Nabataean inscriptions published in Jaussen & Savignac 1909–1914.
LPArab	Arabic inscriptions published in Littmann 1949.
LPNab	Nabataean inscriptions published in Littmann 1914.
M	Site name: Mābiyāt (see Fig. 7).
MS	Site name: Madāʾin Ṣāliḥ (see Fig. 7).
NDGS	Nabataean inscriptions published in Negev 1967.
QN	Site name: Qāᶜ an-Nqayb (see Fig. 7).
RES	*Répertoire d'épigraphie sémitique.* Paris, 1900–1968.
S	Site name: Sakākā (see Fig. 7).
Stiehl	Nabatean inscription published in Stiehl 1970.
ThNUJ	Nabataean inscriptions published in al-Theeb 2002.
ThNUJT	Nabataean inscriptions published in al-Theeb 2005.
UJadh	Site name: Umm Jadhāyidh (see Fig. 7).
ᶜUlā	Site name: al-ᶜUlā (see Fig. 7).
WH	Safaitic inscriptions published in Winnett & Harding 1978.

References

Altheim F. & Stiehl R.

1968. *Die Araber in der alten Welt.* v/1. *Weitere Neufunde — Nordafrika bis zur Einwanderung der Wandalen — Ḏū Nuwās.* Berlin: de Gruyter.

Calvet Y. & Robin R. (eds)

1997. *Arabie heureuse, Arabie déserte. Les antiquités arabiques du musée du Louvre.* (Notes et documents des musées de France, 31). Paris: Éditions de la Réunion des musées nationaux.

Cantineau J.

1930–1932. *Le Nabatéen.* (2 volumes). Paris: Leroux.

Caskel W.

1966. *Ǧamharat an-Nasab. Das genealogische Werk des Hišām ibn Muḥammad al-Kalbī.* (2 volumes). Leiden: Brill.

Cross F.M.

1986. A New Aramaic Stele from Taymāʾ. *Catholic Biblical Quaterly* 48: 387–394.

al-Ghabbān ᶜA.I.

2008. The inscription of Zuhayr, the oldest Islamic inscription (24 AH/AD 644–645), the rise of the Arabic script and the nature of the early Islamic state. (Translation and concluding remarks by R.G. Hoyland). *Arabian Archaeology and Epigraphy* 19: 210–237.

al-Ghul O.

2004. An Early Arabic Inscription from Petra Carrying Diacritic Marks. *Syria* 81: 105–118.

Grimme H.

1936. À propos de quelques graffites du temple de Ramm. *Revue Biblique* 45: 90–95.

Grohmann A.

1932. Aperçu de papyrologie arabe. *Études de papyrologie* 1: 23–95, pls 1–9.

1971. *Arabische Paläographie.* ii. *Das Schriftwesen. Die Lapidarschrift.* (Österreichische Akademie der Wissenschaften, Wien. Philosophisch-historische Klasse. Denkschriften, 94/2). Vienna: Böhlaus.

Gruendler B.
1993. *The Development of the Arabic Scripts. From the Nabatean Era to the First Islamic Century According to Dated Texts*. (Harvard Semitic Studies, 43). Atlanta, GA: Scholars Press.

Healey J.F.
1990–1991. Nabataean to Arabic: Calligraphy and Script Development Among the Pre-Islamic Arabs. In J. Bartlett, D. Wasserstein & D. James (eds), The Role of the Book in the civilisations of the Near East. Proceedings of the Conference held at the Royal Irish Academy and the Chester Beatty Library, Dublin, 29 June – 1 July 1988. *Manuscripts of the Middle East* 5: 41–52.
1993. *The Nabataean Tomb Inscriptions of Madaʾin Salih*. Edited with Introduction, Translation and Commentary. With an Arabic section translated by Dr. Solaiman al-Theeb, Department of Archaeology, King Saud University, Riyadh. (Journal of Semitic Studies Supplement, 1). Oxford: Oxford University Press.
2002. Nabataeo-Arabic: Jaussen-Savignac nab. 17 and 18. Pages 81–90 in J.F. Healey & V. Porter (eds), *Studies on Arabia in Honour of Professor G. Rex Smith*. (Journal of Semitic Studies Supplement, 14). Oxford: Oxford University Press.

Healey J.F. & Smith G.R.
1989. Jaussen-Savignac 17 — The Earliest Dated Arabic Document (A.D. 267). *Atlal* 12: 77–84, pl. 46. [Arabic version pp. 101–110].

Hoyland R.
(this volume). Mount Nebo, Jabal Ramm, and the status of Christian Palestinian Aramaic and Old Arabic in Late Roman Palestine and Arabia. In M.C.A. Macdonald (ed.), *The development of Arabic as a written language.* (Supplement to the Proceedings of the Seminar for Arabian Studies 40). Oxford: Archaeopress.

Ilan T.
2002. *Lexicon of Jewish Names in Late Antiquity.* Part I. *Palestine 330 BCE – 200 CE*. (Texts and Studies in Ancient Judaism, 91). Tübingen: Mohr Siebeck.

Jaussen A. & Savignac R.
1909–1914. *Mission archéologique en Arabie.* i. *De Jérusalem au Hedjaz, Médain Saleh.* ii. *El-ᶜEla, d'Hégra à Teima, Harrah de Tebouk*. (2 volumes). Paris: Leroux/Geuthner.

Jaussen A., Savignac R. & Vincent H.
1905. ᶜAbdeh (4–9-février 1904) (suite). *Revue Biblique Internationale* [N.S.] 2: 74–89, 235–257, pls 6–10.

Littmann E.
1914. *Nabataean Inscriptions from the Southern Ḥaurân*. (Publications of the Princeton University Archaeological Expeditions to Syria in 1904–1905 and 1909. Division IV. Section A). Leiden: Brill.
1949. *Arabic inscriptions.* (Publications of the Princeton University Archaeological Expeditions to Syria in 1904–1905 and 1909, Division IV. Section D). Leiden: Brill.

Macdonald M.C.A.
1986. ABCs and Letter Order in Ancient North Arabian. *Proceedings of the Seminar for Arabian Studies* 16: 101–168.
2003. Languages, Scripts, and the Uses of Writing Among the Nabataeans. Pages 36–56, 264–266 (endnotes), 274–282 (references) in G. Markoe (ed.), *Petra Rediscovered: Lost City of the Nabataeans*. New York: Abrams/Cincinnati, OH: Cincinnati Art Museum.
2008. Old Arabic (Epigraphic). Pages 464–477 in K. Versteeg (ed.). *Encyclopedia of Arabic Language and Linguistics*, iii. Leiden: Brill.
2009*a*. ARNA Nab 17 and the transition from the Nabataean to the Arabic script. Pages 207–240 in W. Arnold, M. Jursa, W.W. Müller & S. Procházka (eds), *Philologisches und Historisches zwischen Anatolien und Sokotra. Analecta Semitica In Memoriam Alexander Sima.* Wiesbaden: Harrassowitz.
2009*b*. A note on new readings in line 1 of the Old Arabic graffito at Jabal Says. *Semitica et Classica* 2: 223.
(forthcoming). On the uses of writing in ancient Arabia and the role of palaeography in studying them.

Milik J.T. & Seyrig H.

1958. Trésor monétaire de Murabbaᶜât. *Revue Numismatique*, 6ᵉ série, 1: 11–26, pls 1–3.

Milik J.T. & Starcky J.

1975. Inscriptions récemment découvertes à Pétra. *Annual of the Department of Antiquities of Jordan* 20: 111–130, pls. 37–47.

al-Muaikil Kh.I. [= al-Muᶜayqil Kh.I.]

1993. Naqshān ᶜarabiyyān mubakkarān min sakākā. *Addarah* 19/3: 112–131.

2002. Pre-Islamic Inscriptions from Sakākā, Saudi Arabia. Pages 157–169 in J.F. Healey & V. Porter (eds), *Studies on Arabia in Honour of Professor G. Rex Smith*. (Journal of Semitic Studies Supplement, 14). Oxford: Oxford University Press.

(forthcoming). [Inscriptions from the region of Sakākā]. *Adumatu*.

al-Muaikil Kh.I. & al-Theeb S.ᶜA. [= al-Muᶜayqil Kh.I. & al-Dhīyīb S.ᶜA.]

1996. *Al-āthār wa-ʾl-kitābāt al-nabaṭiyyah fī minṭaqat al- jawf*. Riyadh: Jāmiᶜat al-Malik Saᶜūd.

al-Muraykhī M.

(in press). Ṭarḥ jadīd ḥawla manshaʾ al-ḥarf al-ᶜarabī wa mawṭinihi al-aṣlī fī ḍawʾi muktashafāt athariyyah jadīdah. *Dirāsāt fī ʾl-āthār* 2.

al-Muraykhī M. & al-Ghabbān ᶜA.I.

2001. Naqsh wāʾil bin al-jazzāz al-tidhkārī al-muʾarrikh bi-ᶜām 410 M. *Silsilat mudāwalāt al-liqāʾ al-ᶜilmī al-sanawī li-l-jamᶜiyyah* 3: 127–153.

al-Najem M. & Macdonald M.C.A.

2009. A new Nabataean inscription from Taymāʾ. *Arabian Archaeology and Epigraphy* 20: 208–217.

Negev A.

1967. New Dated Nabatean Graffiti from the Sinai. *Israel Exploration Journal* 17: 250–255, pl. 48.

1981. Nabatean, Greek and Thamudic Inscriptions from the Wadi Haggag–Jebel Musa Road. *Israel Exploration Journal* 31: 66–71, pls. 7B–10.

1991. *Personal Names in the Nabataean Realm*. (Qedem, 32). Jerusalem: Hebrew University of Jerusalem.

Nehmé L.

2009. Quelques éléments de réflexion sur Hégra et sa région à partir du IIᵉ siècle après J.-C. Pages 37–58 in J. Schiettecatte & Chr.J. Robin (eds), *L'Arabie à la veille de l'Islam. Bilan clinique*. (Orient & Méditerrannée, 3). Paris: De Boccard.

Robin C.J.

2001. Les inscriptions de l'Arabie antique et les études arabes. *Arabica* 48: 509–577.

2006. La réforme de l'écriture arabe à l'époque du califat médinois. *Mélanges de l'Université Saint-Joseph* 59: 319–364.

2008. Les Arabes de Ḥimyar, des « Romains » et des Perses (IIIᵉ–VIᵉ siècles de l'ère chrétienne). *Semitica et Classica* 1: 167–202.

Said S. & al-Hamad M.

2004. Three Short Nabataean Inscriptions from Umm al-Jimāl. *Proceedings of the Seminar for Arabian Studies* 34: 313–318.

Savignac R.

1932. Chronique. Notes de voyage. — Le sanctuaire d'Allat à Iram. *Revue Biblique* 41: 581–597, pls. 16–19.

Sokoloff M.

1990. *A Dictionary of Jewish Palestinian Aramaic*. (Dictionaries of Talmud, Midrash and Targum, 2). Ramat-Gan: Bar Ilan University Press.

Starcky J.

1978. Langue, écriture et inscriptions. Pages 47–52 in F. Baratte (ed.), *Un royaume aux confins du désert: Pétra et la Nabatène*. Catalogue de l'exposition du Muséum de Lyon tenue du 18 novembre 1978 au 28 février 1979. Lyon: Muséum de Lyon.

Stiehl R.
1970. A New Nabatean Inscription. Pages 87–90 in R. Stiehl & H.E. Stier (eds), *Beiträge zur alten Geschichte und deren Nachleben. Festschrift für Franz Altheim zum 6.10.1968*. ii. Berlin: de Gruyter.
al-Theeb S.ᶜA. [=al-Dhīyīb S.ᶜA.]
1993. *Aramaic and Nabataean Inscriptions from North-West Saudi Arabia.* Riyadh: Maktabat al-Malik Fahd al-Waṭaniyyah.
1994*a*. Dirāsah taḥlīliyyah jadīdah li-nuqūsh nabaṭiyyah min mawqiᶜ al-qalᶜah bi-ʾl-jawf bi-ʾl-mamlakah al-ᶜarabiyyah al-saᶜūdiyyah. *Majallat Jāmiᶜat al-Malik Saᶜūd* 6 (Arts 1): 151–194.
1994*b*. Two New Dated Nabataean Inscriptions from al-Jawf. *Journal of Semitic Studies* 39: 33–40.
2002. *Nuqūsh jabal umm jadhāyidh al-nabaṭiyyah*. Riyadh: Maktabat al-Malik Fahd al-Waṭaniyyah.
2005. *Nuqūsh nabaṭiyyah fī ʾl-jawf, al-ᶜulā, taymāʾ, al-mamlakah al-ᶜarabiyyah al-saᶜūdiyyah*. Riyadh: Maktabat al-Malik Fahd al-Waṭaniyyah.
Winnett F.V. & Harding G.L.
1978. *Inscriptions from Fifty Safaitic Cairns*. (Near and Middle East Series, 9). Toronto: University of Toronto Press.
Winnett F.V. & Reed W.L.
1970. *Ancient Records from North Arabia.* (Near and Middle East Series, 6). Toronto: University of Toronto Press.
Yardeni A.
2000. *Textbook of Aramaic, Hebrew and Nabataean Documentary Texts from the Judaean Desert and Related Material. A. The Documents. B. Translation, Palaeography, Concordance*. (2 volumes). Jerusalem: Ben-Zion Dinur Center for Research in Jewish History.
2001. The Decipherment and Restoration of Legal Texts from the Judaean Desert: A Reexamination of *Papyrus Starcky (P. Yadin 36)*. *Scripta Classica Israelitica* 20: 121–137.

Author's address
Laïla Nehmé, CNRS UMR 8167, Orient & Méditerranée, 27, rue Paul Bert, 94204 IVRY CEDEX, France.
e-mail laila.nehme@ivry.cnrs.fr

M.C.A. Macdonald (ed.), *The development of Arabic as a written language*. (Supplement to the Proceedings of the Seminar for Arabian Studies 40). Oxford: Archaeopress, 2010, pp. 89–102.

The evolution of the Arabic script in the period of the Prophet Muḥammad and the Orthodox Caliphs in the light of new inscriptions discovered in the Kingdom of Saudi Arabia

ᶜAlī Ibrāhīm al-Ghabbān

Summary
The letter forms of the Arabic script were already fully formed by the early sixth century AD, and it is reported in the Hadith and other early Islamic sources that writing was greatly encouraged by the Prophet (pbuh) and the Orthodox Caliphs. The calligraphic developments in the script and the improvements, such as the dotting of letters, which we read about in these sources are matched by what we find in the earliest extant Arabic inscriptions, papyri, and manuscripts of the Islamic period. This paper describes the ways in which the use of writing was encouraged from the earliest years of Islam with the consequent development of different forms of the script, and illustrates these from previously published and newly discovered material.

Keywords: Arabic script, writing, inscriptions, Prophet Muhammad, early Islam

In the light of recent epigraphic discoveries and information obtained from historical sources, this paper will discuss the development of the Arabic script during the first half of the first/seventh century, the era of the Prophet Muḥammad, peace be upon him (pbuh), and the Orthodox Caliphs (*al-khulafāᵓ al-rāshidūn*).[1] In some ways, it forms a companion piece to the paper by Laïla Nehmé in this volume, on the development of the Nabataean script into Arabic between the second and fifth centuries AD in the light of the inscriptions recently discovered in the north and north-west of the Arabian Peninsula.

The time frame of the present study is confined to the era of the Prophet (pbuh) and the Orthodox Caliphs, because this period witnessed important developments and reforms of the Arabic scripts and saw the establishment of the fundamental rules which still govern them today.[2]

Early Arabic inscriptions discovered in the Kingdom of Saudi Arabia and Bilād al-Shām and published by a number of scholars in the second half of the twentieth century, show that the Arabic script used in the north-west of the Arabian Peninsula, developed from the Nabataean Aramaic script (al-Jabbūrī 1977: 51–70). This is now confirmed by the texts presented in the above-mentioned paper by Laïla Nehmé, which were discovered recently by the Darb al-Bakrah Survey Project in the regions of al-Jawf and Tabūk in the Kingdom of Saudi Arabia (al-Ghabbān 2007: 13). They include fourteen inscriptions which are dated to the second, third, fourth, and fifth centuries AD and which clearly demonstrate the evolution of each letter from its Nabataean forms into its shape in the earliest Arabic inscriptions at the beginning of the sixth century AD, 100 years before the rise of Islam.

As well as these fourteen dated inscriptions, the Survey discovered a large number of undated memorial inscriptions on the Darb al-Bakrah, an ancient trade route between Ḥegrā (modern Madāᵓin Ṣāliḥ) and Petra which continued in use after the Roman annexation of the Nabataean kingdom in AD 106 right up to the seventh century AD (al-Ghabbān 2007: 22).

It is clear that the evolution from the Nabataean into what is recognizably the Arabic script was completed at least a century before the first revelation received by the Prophet Muḥammad (pbuh), as is shown by the Zebed inscription of AD 512 (Fig. 1), and those of Jabal Usays (AD 528, Fig. 2) and Ḥarrān (AD 568, Fig. 3) (al-Munajjid 1972: 38).[3]

It should be noted that some letters completed their development from their Nabataean to their Arabic form at a relatively early stage. These are *j*, *ḥ*, initial and medial

[1] The author and the editor are most grateful to Robert Hoyland for his considerable help in the preparation of the translation of this paper.

[2] See recently, Robin's treatment of this subject (2006).

[3] On all these inscriptions see Robin 2006, and Hoyland, this volume. On the Jabal Usays graffito and the Ḥarrān inscription see Larcher, this volume, and on the former also Macdonald 2009 and pp. 141–142 here.

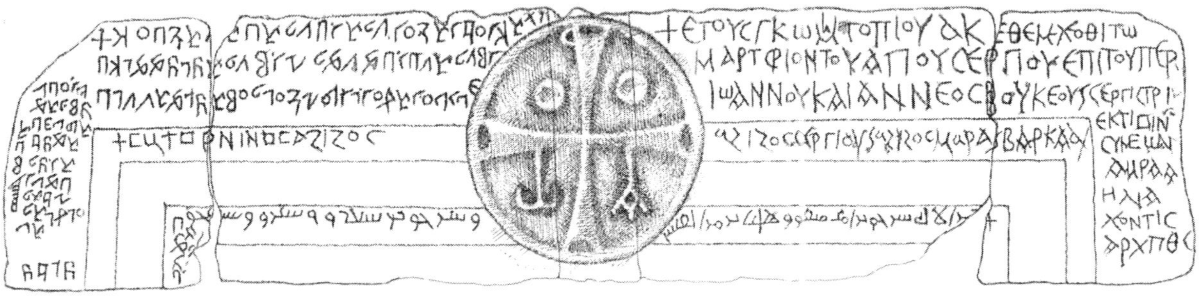

FIGURE 1. *The Zebed inscription (AD 512). (Facsimile by Maria Gorea).*

FIGURE 2. *The graffito of Jabal Usays (AD 528). (Photograph by Michael Macdonald).*

FIGURE 3. *The Ḥarrān inscription (AD 568). (Photograph by Christian Robin).*

FIGURE 4. *JSNab 17 (AD 267). (Photograph by Laïla Nehmé).*

ᶜ, *l*, *alif maqṣūrah*, final "returning" *yā*ʾ, and to a certain extent, initial and final *w*, as can be seen in the inscription JSNab 17, of AD 267, from Ḥegrā/Madāʾin Ṣāliḥ (Fig. 4).[4]

The method of distinguishing letters with the same shape by dots became increasingly common at the very end of the pre-Islamic period, even though all its conventions were not yet fully established.[5] It was called *raqsh*, "embellishment", (al-Asad 1982: 34–41), a word already found with this meaning in pre-Islamic poetry.

> Like the lines of parchment which a scribe *raqasha* ["embellished"/"dotted"] at forenoon.
> (Ṭarafah ibn al-ᶜAbd 1975: 74).

The various types of Arabic script were named after the places in which they developed, such as the *Anbārī* and *Ḥīrī* scripts which were named after the towns of al-Anbār and al-Ḥīrah in Iraq. The former was rounded

[4] On this inscription see Appendix 2 of Nehmé, this volume.

[5] On the use of dots in Nabataean and "transitional" inscriptions, see Nehmé, this volume.

(*mudawwar*) and the latter triangular (*muthallath*). In the Ḥijāz, the best-known script was "the composite" (*taʾim*), which combined features of the rounded and the angular letter shapes. It was later, after the appearance of Islam, called *Makkī* or *Madanī* or *Ḥijāzī* (Ibn al-Nadīm 1978: 8, 9).

The encouragement of writing by the Prophet Muḥammad (pbuh)

When Islam emerged in the early seventh century AD, writing was considered to be a vital tool in spreading the religion. Indeed, it was at its very heart, as demonstrated by the first revelation received by the Prophet (pbuh):

> Read in the name of thy Lord and Creator, Who created man from a (mere) clot of congealed blood. Read! And thy Lord is Most Bountiful. He who taught (the use of) the pen ... (Qurʾān 96: 1–4)

Thus, God's first instructions to the Prophet (pbuh) were "to read" and "to write" as the prime means of spreading the religion, a method which the Prophet (pbuh) used extensively.

Reading and writing are a means of widening understanding, enriching dialogue, and deepening knowledge. Islam put literacy to all these uses and this ultimately led to the introduction of writing into all aspects of life. The miracle of the illiteracy of the Prophet (pbuh) was counted in his favour and it was considered to give credibility to his message. He adopted several ways of encouraging the teaching of writing. From the very first, he permitted the writing of the revelations and these were secretly exchanged among the earliest believers in Mecca, as is shown by the story of the conversion of ᶜUmar ibn al-Khaṭṭāb. He found his sister and her husband reciting verses of the Qurʾān which had been written down. He took the writing from them, read it, was astonished and immediately became a Muslim (Ibn Hishām 1978, ii: 289). Another example of the use of writing can be found in the agreements between the Muslims and the unbelievers in Mecca — including the treaty of Ḥudaybiyyah — which were written down and kept in the Kaᶜbah (Ibn Hishām 1978, ii: 218).

After the Prophet Muḥammad (pbuh) emigrated to Medina and established the Islamic state there, writing became essential for government administration, in addition to its importance in recording the revelations and disseminating the religion. The first known official document to be written was the Constitution of Medina, which specified communal rights and ordered the relations between the communities there. The encouragement of writing by the Prophet (pbuh) can be seen by his release of Qurayshī prisoners on condition that each taught writing to ten boys of Medina (Ibn Saᶜd 1987, ii: 22).

He also encouraged the teaching of writing to women, including his wife, Umm al-muʾminīn ("mother of the faithful"), Ḥafṣah bint ᶜUmar, and others of the female *ṣaḥābah* ("Companions"), who were taught by Shifāʾ bint ᶜAbd Allāh al-Qurayshiyyah (Ibn al-Athīr 1996, vii: 177, no. 7031). He dictated the revelations to up to forty scribes, including ᶜUthmān ibn ᶜAffān and ᶜAlī ibn Abī Ṭālib, important men who acted as scribes when Ubayy ibn Kaᶜb, Zayd ibn Thābit, and others were away. The Prophet (pbuh) dictated religious and general matters to others, such as Khālid ibn Saᶜīd ibn al-ᶜĀṣ and Muᶜāwiyah ibn Abī Sufyān. Some scribes, such as Mughīrah ibn Shuᶜbah and al-Ḥuṣayn ibn 'Umayr, used to record loans and contracts, while others, such as ᶜAbd Allāh ibn Arqam ibn ᶜAbd Yaghawth and ᶜAlāʾ ibn ᶜUqbah, specialized in recording charitable gifts (*ṣadaqāt*), tribes and their water sources; or, like Ḥudhayfah ibn al-Yamān, the crops of the Ḥijāz; or, like Muᶜayqīb ibn Abī Fāṭimah, the share of booty belonging to the Prophet (pbuh); or they were assigned to write letters to kings, or to reply to their letters, or to make translations into Persian, or Greek, or Ethiopic, like Zayd ibn Thābit (al-Ḥamad 2003: 49).

In this way, Islam greatly increased the use of writing. Indeed, the Qurʾān orders the writing-down of loans and transactions among Muslims (e.g. 2: 282) and between them and non-Muslims (e.g. 9: 1), thus making writing a daily practice at the time of the Prophet (pbuh).

The writing-down of the Qurʾān during the lifetime of the Prophet Muḥammad (pbuh)

Whenever the Prophet (pbuh) received a revelation, he would first memorize it himself and then dictate it to his Companions (*ṣaḥābah*) and tell them into which *sūrah* and after which *āyah* (verse) this new revelation should be inserted. The written Qurʾān would have been kept at the house of the Prophet (pbuh), but some of his scribes made copies for themselves of the verses they wrote down (Ibn Ḥajar 1960, ix: 12). The verses of the Qurʾān were written on palm-leaf stalks, thin flat stones, pieces of cloth, scraps of leather, and the shoulder blades and ribs of animals. The complete Qurʾān, along with the correct sequence of the verses, existed during the lifetime of the Prophet (pbuh), but was not compiled in one book (*muṣḥaf*), because the Prophet (pbuh) was expecting additional verses or deletions and his Companions preferred to memorize it rather than write it down. Indeed, by his death, most of them had already memorized the whole or part of Qurʾān as well as writing down all or part of it. Thus, the complete Qurʾān was both memorized and written down in the lifetime of the Prophet (pbuh).

The contribution of the Prophet Muḥammad (pbuh) to the development of writing and the differentiation of letters of similar form by the use of dots

As stated above, the letter forms of the Arabic script had completed their evolution at least a century before the rise of Islam, and dots to distinguish between letters of similar shape were introduced during the same period. It has also been mentioned that the calligraphic style known as "the composite" (*al-taʾim*), which combined elements of the rounded script of al-Anbār and angular style of al-Ḥīrah, was favoured in the Ḥijāz area, particularly in Mecca, Medina, and Ṭāʾif. The *Ḥijāzī* form of the script was the one used in Mecca and Medina at the time of the Prophet (pbuh) and would have been the one used to write down the Qurʾān as well as official correspondence and the transactions of daily life in the lifetime of the Prophet (pbuh) (al-Ḥamad 2003: 54). The *Ḥijāzī* script was characterized by the use of dots on letters when deemed necessary, though it was not yet the norm, and the dots were added, or not, at the discretion of each scribe. The Prophet (pbuh) used to direct his amanuenses to give due regard to dotting and he is quoted as saying, "in cases of disagreement regarding *yāʾ* [which has two dots below the letter] and *tāʾ* [which has two dots above the letter] write *yāʾ*" (Ibn al-Athīr 1996, ix: 12). The Prophet (pbuh) also gave the instruction "Fix knowledge by writing it down" (al-Jabbūrī 1977: 156; see also Cook 1997), and to Muʿāwiyah ibn Abī Sufyān he said: *urqush kitābaka*, i.e. "place dots on the letters that need them".[6] Dots are believed to have been used in the writing of the Qurʾān at the time of the Prophet (pbuh), since al-Farrāʾ reports one of the second-generation Muslims (*al-tābiʿūn*) as saying:

> I wrote on a stone *nunsiruhā* and *lam yatasanna*. I looked to Zayd ibn Thābit and he put dots on the *shīn* and the *zāy*, four in all [making the word read *nunshizuhā*], and he added a *hāʾ* [on the end of] *yatasanna* (al-Farrāʾ 1955, i: 172).[7]

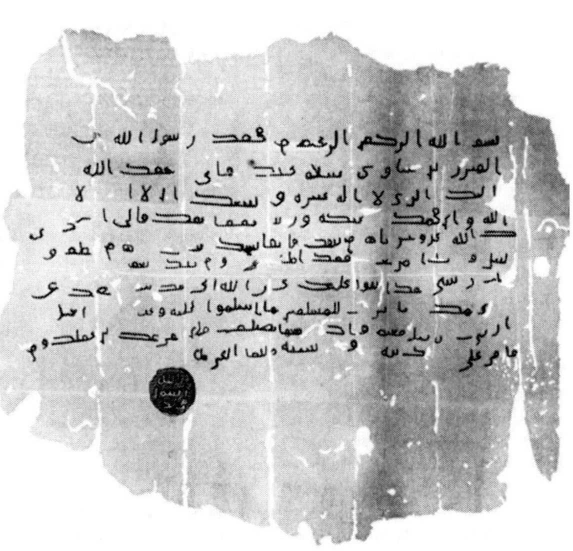

Figure 5. *The letter said to have been sent by the Prophet (pbuh) to al-Mundhir Ibn Sāwī of Bahrain. (From Hamidullah 1985–1986: 111).*

The reason why the complete Qurʾān was written down without dots under the caliph ʿUthmān ibn ʿAffān will be explained below.

It should be noted, however, that no surviving written materials or inscriptions can be firmly dated to the lifetime of the Prophet (pbuh). Although they are mentioned in the historical sources, the "extant" letters said to have been sent by the Prophet (pbuh) to al-Mundhir Ibn Sāwī of Bahrain (Fig. 5) or to al-Muqawqis, ruler of Egypt, etc. cannot be proved to be genuine. Despite the fact that they are written on parchment — an organic material that can be dated by ^{14}C — no laboratory test has ever been made to determine their dates. It is also worth noting that these letters could have been produced, or modified, in the early Islamic era. Some of them contain spelling and writing mistakes, which could have resulted from attempts to rewrite faded and unclear parts. It has also been noted that the letter alleged to have been sent from the Prophet (pbuh) to Ibn Sāwī was written in the *Ḥijāzī* script, which was in use during the first century of the Hijrah (al-Munajjid 1972: 34) (see Fig. 5).

The Prophet (pbuh) considered the dissemination of writing among Muslims as a divine command and part of his mission. He exerted his maximum efforts to realize this while he was in Mecca and Medina and God Almighty ensured the success of the Prophet (pbuh) in transforming an uncultured and illiterate people into a community using reading and writing for its advancement and civilization. In the words of God:

[6] See also on this subject Cook 1997.

[7] The reference is to Qurʾān 2.259, which contains the words *nunshizuhā* and *lam yatasannah*; the former word is ambiguous without dots, so Zayd added dots to make it clear. [Robert Hoyland].

> It is He Who has sent amongst the unlettered an apostle from amongst themselves, to rehearse to them His Signs, to sanctify them, and to instruct them in Scripture (*al-kitāb*) and Wisdom. (Qurᵓān 62: 2).

This verse was revealed to the Prophet (pbuh) in Medina, confirming his success in realizing the divine command he had received. Ibn ᶜAbbās interpreted the word *al-kitāb* in the verse as meaning not only the Qurᵓān, but writing in general (*al-ḫaṭṭ wa-ᵓl-qalam*, "the script and the pen"), because use of the script had become a legal requirement for the Arabs once they were ordered to put things down in writing (al-Qurtubī 1952, xviii: 92).

The development of writing in the period of the Orthodox Caliphs (*al-khulafāᵓ al-rāshidūn*)

The time of the Prophet Muḥammad (pbuh) witnessed the completion of the rules for distinguishing between letters with identical shapes by the use of dots, and a remarkable expansion in the teaching of writing and its use in the religious, official, and daily life of the Muslim community. It was followed by the era of the Orthodox Caliphs (11–40/633–661) when focus moved to improving calligraphy and developing the styles and shapes of the letters. The improvement of calligraphy began with the introduction of the "*alif* of prolongation" (*al-alif al-madd*, representing [a:]), as well as introducing some decorative additions to the letters. The collection of the Qurᵓān under the first caliph, Abū Bakr, and the production of several copies, required the services of a large number of scribes and calligraphers and this must have promoted writing and the art of calligraphy at this early stage of Islamic history.

The collection of the Qurᵓān under the caliph Abū Bakr

The writing-down of the Qurᵓān began and was completed during the lifetime of the Prophet (pbuh) and was reviewed and checked in two stages: firstly, at the time of the recording of the verses which the angel Jibrīl brought down, and secondly, during the review of the sections when they were transcribed and arranged (al-Ṭabarānī 1984, v: 142). However, the Qurᵓān was not collected in one text or in an organized form until the death of the Prophet (pbuh); rather it existed on scattered tablets, parchments, and palm-leaf stalks (Ibn Ḥajar 1960, ix: 12). During his caliphate (11–13/632–634), Abū Bakr was compelled to fight apostates, with the result that many of the Companions were martyred, among them about fifty persons who had memorized the Qurᵓān (Ibn Khayyāṭ 1967, i: 79). Because of this, the Companions of the Prophet (pbuh) feared that the text of the Qurᵓān might be lost or forgotten and decided to compile the Revelations into one unified collection.[8] Accordingly, the caliph Abū Bakr ordered Zayd ibn Thābit, who had persuaded the Companions to write down the Revelations during the lifetime of the Prophet (pbuh), to collect them, and Zayd finished his assignment successfully (al-Bukhārī [n.d.], vi: 89, 225). This is considered one of the most important events in the development of writing and calligraphy in the period of the Orthodox Caliphs. This assignment was not undertaken by Zayd ibn Thābit alone — though he shouldered the major responsibility for it — but, rather, it was a team effort in which many of the Companions and followers took part. This work did not consist in writing down the Qurᵓān for the first time, but rather was a recopying of the Quranic text, which had been written down originally in the presence of the Prophet (pbuh). To ensure accuracy, Zayd also used portions of the Qurᵓān which had been memorized, though he would only accept these if they were confirmed by two other memorizers (al-Suyūṭī 1976, i: 66). Those assigned to the compiling of the Qurᵓān under the direct supervision of Zayd ibn Thābit were divided into two teams: one dictating and the other writing. ᶜUmar ibn al-Khaṭṭāb was a member of the latter team and was quoted as saying "Dictation of the Qurᵓān (*maṣāḥif*) should only be undertaken by boys of [the tribe of] Thaqīf." Members of the Thaqīf tribe were well known for their abilities in writing. The compilation and writing took up a considerable part of the year 12/634. The Qurᵓān was written on carefully arranged folios made from pieces of hide (al-Ṭabarī 1967, i: 26) and, so it was said, of Egyptian papyrus (Ibn Ḥajar 1960, ix: 16). From then on, the written Qurᵓān was kept between two covers. ᶜAli ibn Abī Ṭālib was quoted as saying "May God Almighty bless Abū Bakr, for he was the first to collect the Qurᵓān between two covers" (al-Sijistānī 1936–: 5). The folios bearing the written Qurᵓān were kept in the custody of Abū Bakr, then the caliph ᶜUmar ibn al-Khaṭṭāb and finally upon the latter's death, it was taken to his daughter Ḥafṣah bint ᶜUmar, "mother of the faithful". Later, the caliph ᶜUthmān ibn ᶜAffān had a unified version made and all others were burnt.

The volume (*muṣḥaf*) of the Qurᵓān written during the reign of the caliph Abū Bakr may well have been in the

[8] It should be noted, however, that although many of the Companions who had memorized the Qurᵓān were martyred, the complete Qurᵓān in writing had already been available since the time of the Prophet (pbuh).

Ḥijāzī script, which was also used in the lifetime of the Prophet Muḥammad (pbuh), and it is possible that dots may have been employed.

The collection of the Qurʾān under the caliph ʿUthmān

The Prophet (pbuh) received the Qurʾān in different Arabic dialects and he allowed it to be recited in these dialects, with differing pronunciations. With the spread of Islam to many areas, the Companions went out to these areas to teach people the Qurʾān in the form in which they had learnt it. These differences led to many variations in the text between one place and another, since there was no unified *muṣḥaf*. When the caliph ʿUthmān learnt of the differences in reciting the Qurʾān, he consulted the Companions and decided to transcribe the copy of the Qurʾān that had been compiled in the time of Abū Bakr, and write it in the dialect of the Quraysh, as recited by the Prophet (pbuh). When the scribes had prepared several copies of the text compiled under Abū Bakr, one copy was sent to each city under Muslim rule. All the versions that did not accord with this Uthmanic recension were burnt (al-Bukhārī [n.d.], vi: 226).

In order to accommodate all possible readings of some words it was decided to remove all dots. This contributed to the solution of the core problem resulting from the decision to transcribe the Qurʾān into one unified *muṣḥaf*, in order to make things easier for the people and to avoid further differences. For instance, as noted above, the Arabic word *nunshizuhā* could also be read *nunshiruhā* if the letter *zāy* was not dotted, and indeed both readings were heard from the Prophet (pbuh) in respect of the Quranic verse (2: 259): "Look at the bones, how we put them together (*nunshizuhā*) / lay them out (*nunshiruhā*), and clothe them with flesh."

Initially, the caliph ʿUthmān ibn ʿAffān entrusted Zayd ibn Thābit, ʿAbd Allāh ibn al-Zubayr, Saʿīd ibn al-ʿĀṣ, and ʿAbd al-Raḥmān ibn al-Ḥārith ibn Hishām with this task and when they needed others to help them, the caliph added eight more people from the Quraysh and from the *anṣār*, including Ubayy ibn Kaʿb (al-Ḥamad 2003: 64, 65). It seems that this group was in charge of overseeing the process of copying and checking the verses, while the actual dictation and writing were carried out by others. In this regard it is reported that ʿUthmān ibn ʿAffān said, "select those who dictate from the Hudhayl [tribe] and the scribes from the Thaqif [tribe]" (al-Dānī 1940: 9) The script used to write the *muṣḥaf* must have been the *Ḥijāzī*, which developed in a remarkable way during the caliphate of ʿUthmān, particularly after the introduction of the *mashq*, i.e. the elongation of letters.

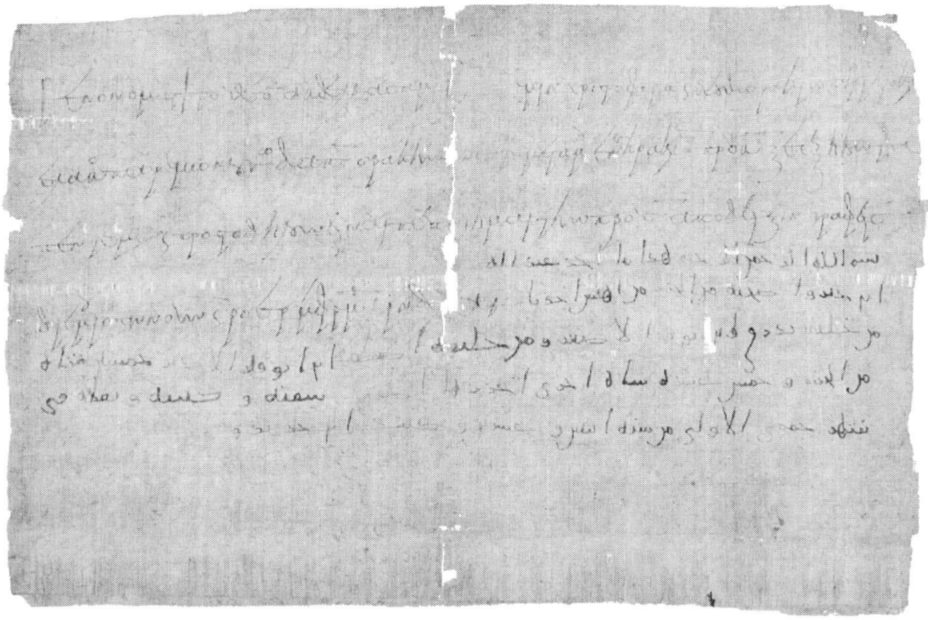

FIGURE 6. *The Ahnās papyrus (22/643), PERF 558 recto.*
(From www.onb.ac.at/sammlungen/papyrus/papyrus_bestandsrecherche.htm#).

Some historical reports indicate that four *maṣāḥif* were copied during the caliphate of ᶜUthmān, while others refer to seven: one was retained in Medina (al-Dānī 1940: 9) and six were distributed to Mecca, Syria, Yemen, Bahrain, Basra, and Kufa. Whichever report is correct, this version of the text was transcribed in the same order and style, by Muslims over a wide area (al-Ḥamad 2003: 66).

The insertion of the medial *alif* of prolongation in proper nouns and other words

Medial [a:] remains unrepresented in both proper nouns and other words in all the pre-Islamic Arabic inscriptions. For instance, the personal name Amat-Manāf appears in the Zebed inscription (AD 512, Fig. 1) as *ʾmtmnfw*, the personal name al-Ḥārith is written in the Jabal Usays graffito (AD 528, Fig. 2) as *ʾlḥrth*, and the name Sharāḥīl as *shrḥyl* in the Ḥarrān inscription of AD 568 (Fig. 3),

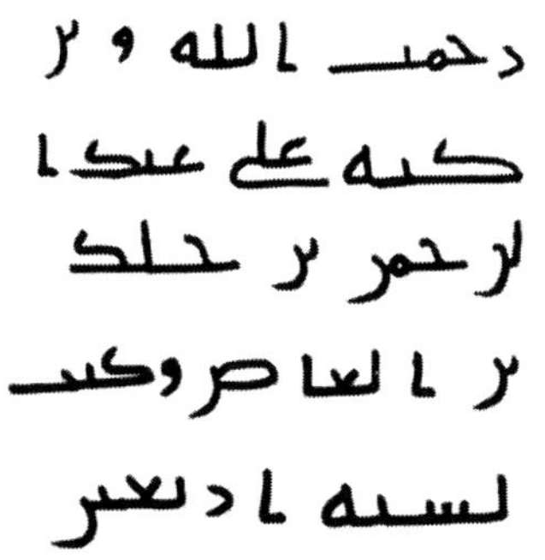

Figure 7. *The Arabic inscription from Bāthā in the Wādī ʾl-Shāmiyyah, of 40/660–661. (From Sharafaddin 1977: 60–70, pl. 49B).*

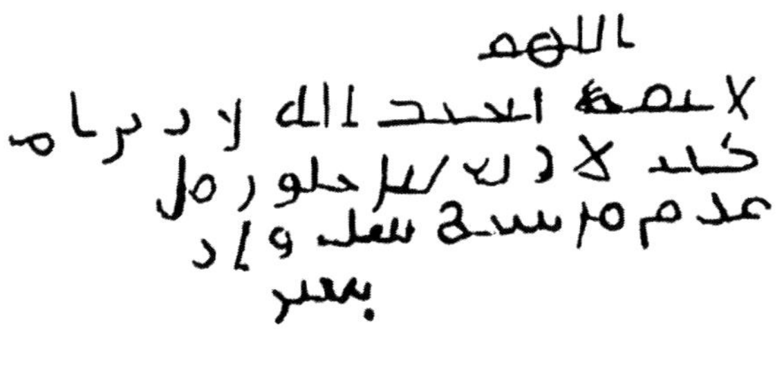

Figure 8. *The Arabic inscription from Wādī Sabīl, near Najrān, of 46/666. (From Grohmann 1962: 124, no. Z 202, pl. 23).*

while the phrase *bi-ᶜām* is written *bᶜm* in the same inscription. The first historically documented use of *alif* to represent medial [a:] in the Arabic script is found in the Ahnās papyrus (PERF 558) of 22/643 (Fig. 6), i.e. in the reign of the second caliph, ᶜUmar ibn al-Khaṭṭāb.[9] There, *alif* representing medial [a:] is found on two occasions, both in the word *shāh* (lines 4 and 5). Chronologically, the next example is in the Bāthā inscription found in Wādī ʾl-Shāmiyyah (Sharafaddin 1977: 60–70, pl. 49A, B), which is by a certain ᶜAbd al-Raḥmān ibn Khālid ibn al-ᶜĀṣ dated 40/660–661 (Fig. 7), thus coinciding with the end of the period of the Orthodox Caliphs. Here, the [a:] in the name al-ᶜĀṣ is spelt with an *alif*, even though the other two names do not have this feature. This shows that the use of the *alif* of prolongation was not considered essential at this time, but was only employed to highlight an important word, such as *shāh* ("sheep") in the Ahnās papyrus which was a receipt for sixty-five sheep, or to avoid the misreading of a name, like *al-ᶜĀṣ* in the Bāthā inscription, or to ensure the correct reading of an unusual name such as *Dīrām* at the end of line 1 of the inscription from Wādī Sabīl, near Najrān, of 46/666 (Fig. 8) (Grohmann 1962: 124, no. Z 202, pl. 23).

The introduction of the *mashq* style in Arabic calligraphy

The *mashq* style, also known as "elongation", was used throughout the period of the Orthodox Caliphs and was characterized by the extension of individual letters on the line to create decorative shapes, and to help ensure the evenness of the ends of lines. There was considerable disagreement over whether it was a good or bad style. The caliph ᶜUmar ibn al-Khaṭṭāb is reported to have said "the best calligraphy is the clearest" and that he hated the use of *mashq* in copying the Qurʾān (Ibn Manẓūr 1999, xiii: 116; al-Ṣūlī 1922: 56). In any case, there are no archaeological materials or manuscripts attributable to the period of the Orthodox Caliphs written in this style, and there are no clear examples of *mashq* in the letters attributed to the Prophet (pbuh) (e.g. Fig. 5).

At present, the earliest clear example of *mashq* is in an undated inscription on Jabal Salaᶜ in Medina, the date of which is disputed. Hamidullah dated it to the fifth year of the Hijrah (AD 626–627) and linked it to the campaign of the Khandaq because it was found along the route where the battle at Medina occurred (1939: 428). Hamidullah read the inscription as follows:

[9] See von Karabacek 1894, Jones 1998, and Larcher, this volume.

Figure 9. *The Arabic inscription from Jabal Salaᶜ, in Medina. (From Hamidullah 1939: pl. 8).*

> *amsā wa-aṣbaḥa ᶜUmar*
> *wa-Abū Bakr yatawaddaᶜāni* (or *yatūbāni* ?, or *yataḍarraᶜāni* ?)
> *ilā Allāh min kull*
> *mā yakrahu* (1939: 434).

and translated it:

> Night and day ᶜUmar and Abū-Bakr take shelter (?) with [or "repented before", or "made entreaty to", (ed.)] God from everything unpleasant.

Hamidullah's reading and his dating of the inscription have frequently been cited in later treatments (e.g. al-Munajjid 1972: 31). However, I believe that the correct reading is:

> *amsā wa-aṣbaḥa*
> *Abū Bakr yatūbu*
> *ilā Allāh min kull*
> *mā yakrahu*

> Abū Bakr spent all night till morning repenting, asking God to forgive him for anything that He dislikes.

I believe that what Hamidullah read as the name "ᶜUmar" (at the end of line 1) and the *alif* and *nūn* of the word he read *yatūbāni* (at the end of line 2) are simply scratches, cracks, and the remains of adjacent inscriptions. Moreover, it is not plausible to have ᶜUmar preceding Abū Bakr. I believe that the Jabal Salaᶜ inscription dates to the period of Abū Bakr, i.e. 11–13/632–634.

It should be noted that in the word *yatūb* in this inscription, there are two dots arranged vertically over the *t*, and one dot beneath the *b*. It also seems likely that the *b* of *aṣbaḥa* had a dot beneath it, although this is less certain. The dots are rounded and are consistent in style and usage with those in the papyri and inscriptions of the period of the Orthodox Caliphs (al-Ghabbān 2008: 226). The letter forms also match those of the scripts used at that period, particularly the *alif maqṣūrah* in the words *amsā* and *ilā*, the initial *alif* and the final *ḥāʾ* in *aṣbaḥa*, and the *rāʾ* in *Abū Bakr* and *yakrahu*. All this supports the proposition that this inscription dates to the caliphate of Abū Bakr, despite the fact that it is in a developed form of the script using the *mashq* style. I consider this important inscription to be substantial evidence for the advanced state of Arabic calligraphy in the period of the Orthodox caliphs.

The *mashq* style can also be seen in the first Zuhayr inscription dated 24/645–646 (al-Ghabban 2008) (Fig. 10) in the backward curve of the final *y* of *tuwuffiya* which runs back below the other letters of the word, a feature which can also be observed in the *alif maqṣūrah* of the word *mawlā* in the second Zuhayr inscription (Fig. 11),[10] and on the leaves of the copy of the Qurʾān which is thought to have belonged to the Caliph ᶜUthmān ibn ᶜAffān.

The concern for rounding the tails of final forms of letters descending below the line

As well as using the *mashq* style, it is noticeable that scribes in the period of the Orthodox caliphs also rounded the final forms of letters which extend below the line such as *alif maqṣūrah* and *yāʾ* and the almost semicircular *nūn*, *sīn*, and *ṣād*. Typical examples of the final forms of these letters can be seen in the *alif maqṣūrah*, which occurs twice in the Jabal Salaᶜ inscription in the words *amsā* and *ilā* (Fig. 9), as well as in the Ahnās papyrus in the words *jumādā al-ūlā*, and *ukhrā* (Fig. 6). In the same papyrus, final *nūn* is written twelve times with a large rounded shape but also four times as a vertical stroke in the word *ibn*. Final *nūn* occurs with the large rounded shape, twice in the first Zuhayr inscription in the words *zaman* and *ᶜishrīn* (Fig. 10); three times on the Aswān tombstone (Hawary 1930) dated 31/652 in *ᶜabd al-raḥmān*, *amīn*, and *thalāthīn* (Fig. 12) (al-Munajjid 1972: 40); and four times in the Bāthā inscription found in Wādī ʾl-Shāmiyyah (Sharafaddin 1977: 69–70, pl. 49/A, B) (Fig. 7). Similar to this is the tail of final *sīn* in the word *khams* in the Ahnās papyrus (Fig. 6) and the tail of final *ṣād* in the name *al-ᶜĀṣ* in the Bāthā inscription (Fig. 7).

[10] On the Zuhayr inscriptions, see Ghabban 2008.

Figure 10. *The first "Zuhayr" Arabic inscription of 24/645–646 near Qāᶜ al-Muᶜtadil, south of Madāʾin Ṣāliḥ. (Photograph by the author).*

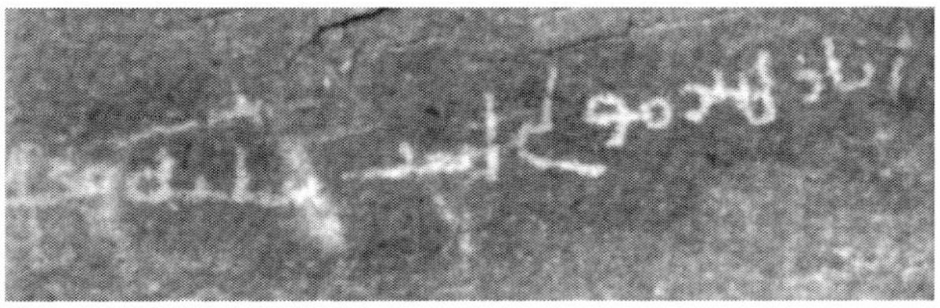

Figure 11. *The second "Zuhayr" Arabic inscription from near Qāᶜ al-Muᶜtadil, south of Madāʾin Ṣāliḥ. (Photograph by the author).*

Figure 12. *A tombstone from Aswān dated 31/652. (From El-Hawary 1930: pl. 3).*

Figure 13. *An Arabic inscription in Mecca dated to 80/699–700. (From Rāshid 1995: 160, Appendix no. 1).*

Figure 14. *An Arabic inscription in Mecca dated to the first century AH. (From Rāshid 1995: 164, Appendix no. 2).*

The downward extension of the tails of some letters

In a further development of the phenomenon of extending the letters across the page and lengthening them, scribes in the period of the Orthodox Caliphs practised extending the tails of final letters descending vertically below the line of writing, elongating their ends with decorative additions. This is particularly obvious in the letter *qāf*, but is sometimes also applied to *fāʾ*, *nūn*, and *yāʾ*. The earliest example is in line 3 of the Ahnās papyrus (22/645) in the name *tadhraq* (Fig. 6), but it is also found in the final *yāʾ* of an inscription discovered recently on the Syrian Hajj route, which can be dated to the second half of the first century AH (al-Kilābī 2009: 139); and in the final *qāf* and *nūn* in the ᶜUthmān ibn Wahrān inscription from Mecca of 80/699–700 (Fig. 13 and cf. Fig. 14) (al-Ḥārithī 2009: 489).

Equalizing the apices/tips of *alif* and *lām*

Another characteristic of the *mashq* style was the lengthening of the letters *alif* and *lām* to the same height, so that their tips were on the same level, in order to give the whole line an aesthetically pleasing regularity of height. This practice seems to have begun with the writing of these letters in the word *Allāh* to make it stand out in the text. Later, it was applied to all examples of these letters and then to the upper tips of *kāf*, *dāl*, *dhāl*, *ṭāʾ*, and *ẓāʾ*. Building on this, other decorative styles were introduced, such as broadening (*taᶜrīḍ*), which consists of giving triangular tips to the stems of these letters, a practice which then developed into the form of decoration known as foliation (*tawrīq*).

The lengthening and equalization of the stems of *alif* and *lām* can already be found in inscriptions of the period of the Orthodox Caliphs, indicating that this practice, which appears in a developed form in the second half of the first century AH, must have originated before or during the first half. Thus, all the examples of *alif* and *lām* in the Ahnās papyrus show this characteristic (Fig. 6), as do all examples of *Allāh* in the first Zuhayr inscription (Fig. 10). Similarly in the Aswān tombstone (Fig. 12), despite the rather crude quality of the script, the tip of the *kāf* is level with the tips of the two examples of *alif* in its vicinity in the sentence *wa-adkhalahu fī raḥmah minka wa-iyyanā maᶜahu*. Other examples can be seen in the word *Allāh* in the Jabal Salaᶜ inscription (Fig. 9) and in the pages believed to be from the caliph ᶜUthmān's copy of the Qurʾān.

Sigla

JSNab	Nabataean inscriptions in Jaussen & Savignac 1909–1922.
PERF	Papyri of the Archduke Rainer. See von Karabacek 1894.

References

al-Asad N.
1982. *Maṣādir al-shiᶜr al-jāhilī wa-qīmatuhā ʾl-taʾrīkhiyyah*. Cairo. [Publisher unknown].

al-Bukhārī, Muḥammad b. Ismāᶜīl/[Editor unknown].
[n.d.]. *Jāmiᶜ al-ṣaḥīḥ*. Cairo. [Publisher unknown].

Cook M.
1997. The opponents of the writing of Tradition in early Islam. *Arabica* 44: 437–530.

al-Dānī, Abū ᶜAmr ᶜUthmān b. Saᶜīd/[Editor unknown]
1940. *Kitāb al-Muqniᶜ fī maᶜrifat rasm maṣāḥif al-amṣār*. Damascus. [Publisher unknown].

al-Farrāʾ, Abū Zakariyyāʾ Yaḥyā b. Ziyād/[Editor unknown]
1955. *Maᶜānī ʾl-qurʾān*. Cairo: Dār al-Kutub al-Maṣriyyah.

al-Ghabbān ᶜA.I.
2007. Le Darb al-Bakra: découverte d'une nouvelle branche sur la route commerciale antique, entre al-Ḥigr (Arabie Saᶜūdite) et Pétra (Jordanie). *Comptes rendus de l'Académie des Inscriptions et Belles-Lettres*: 9–24.
2008. The inscription of Zuhayr, the oldest Islamic inscription (24 AH/AD 644–645), the rise of the Arabic script and the nature of the early Islamic state. (Translation and concluding remarks by R.G. Hoyland). *Arabian Archaeology and Epigraphy* 19: 210–237.

Grohmann A.
1962. *Arabic Inscriptions.* Expédition Philby-Ryckmans-Lippens en Arabie. IIᵉ Partie: Textes épigraphiques. Tome 1. (Bibliothèque du Muséon, 50). Louvain: Institut Orientaliste.

Ḥamad Gh.Q.
2003. *Muḥāḍarāt fī ᶜulūm al-qurʾān*. Amman. [Publisher unknown].

Hamidullah M.
1939. Some Arabic inscriptions of Medinah of the early years of the Hijrah. *Islamic Culture* 13: 427–439, pls 2–10.
1985–1986. *Six originaux des lettres diplomatiques du Prophète de l'Islam, avec une introduction à l'origine de l'écriture arabe*. Paris: Tougui.

Ḥārithī N.
2009. *Al-āthār al-islāmiyyah fī makka*. Riyadh. [Publisher unknown].

El-Hawary H.
1930. The Most Ancient Islamic Monument known dated A.H. 31 (A.D. 652) from the time of the third Calif ᶜUthman. *Journal of the Royal Asiatic Society of Great Britain and Ireland*: 321–333.

Hoyland R.
(this volume). Mount Nebo, Jabal Ramm, and the status of Christian Palestinian Aramaic and Old Arabic in Late Roman Palestine and Arabia. Pages 29–46 in M.C.A. Macdonald (ed.), *The development of Arabic as a written language*. (Supplement to the Proceedings of the Seminar for Arabian Studies 40). Oxford: Archaeopress.

Ibn al-Athīr, ᶜIzz al-Dīn Abū ʾl-Ḥasan ᶜAlī/ed. ᶜA.A. al-Rifāᶜ ī
1996. *Usd al-ghābah fī maᶜrifat al-ṣaḥābah*. Beirut: Dār iḥyāʾ al-turāth al-ᶜarabī.

Ibn Ḥajar, Shihāb al-Dīn Abū ʾl-Faḍl Aḥmad b. Nūr al-Dīn al-ᶜAsqalānī/[Editor unknown]
1960. *Fatḥ al-bārī sharḥ ṣaḥīḥ al-bukhārī*. Cairo. [Publisher unknown].

Ibn Hishām, Abū Muḥammad ᶜAbd al-Malik/[Editor unknown]
1978. *Al-sīrah al-nabawiyyah*. Beirut. [Publisher unknown].

Ibn Khayyāṭ al-ᶜUṣfurī, Khalīfah/[Editor unknown]
1967. *Al-taʾrīkh*. Damascus. [Publisher unknown].

Ibn Manẓūr, Muḥammad b. Mukarram/[Editor unknown]
1999. *Lisān al-ᶜarab*. Beirut. [Publisher unknown].

Ibn al-Nadīm, Abū ʾl-Faraj Muḥammad b. Isḥāq/[Editor unknown]
1978. *Kitāb al-Fihrist*. Beirut: Dār al-Maᶜrifah.

Ibn Saᶜd, Abū ᶜAbd Allāh Muḥammad/[Editor unknown]
1987. *Kitāb al-ṭabaqāt al-kabīr*. Beirut: Dār Maṣādir.

al-Jabbūrī S.Y.
1977. *Aṣl al-khaṭṭ ᶜarabī wa-taṭawwurhu ḥattā nihāyat al-aṣr al-umawī*. Baghdad. [Publisher unknown].

Jaussen A. & Savignac M.R.
1909–1922. *Mission archéologique en Arabie*. (5 volumes). Paris: Leroux/Geuthner.

Jones A.
1998. The Dotting of a Script and the dating of an era: the strange neglect of PERF 558. *Islamic Culture* 72: 95–103.

Karabacek J. von
1894. *Papyrus Erzherzog Rainer: Führer durch die Ausstellung*. (Arabische Abteilung). Vienna: Selbstverlag der Sammlung.

al-Kilābī H.
2009. *Al-nuqūsh al-islāmiyyah*. Riyadh. [Publisher unknown].

Larcher P.
(this volume). In search of a standard. Dialect variation and New Arabic features in the oldest Arabic written documents. Pages 103–112 in M.C.A. Macdonald (ed.), *The development of Arabic as a written language*. (Supplement to the Proceedings of the Seminar for Arabian Studies 40). Oxford: Archaeopress.

Macdonald M.C.A.
2009. A note on new readings in line 1 of the Old Arabic graffito at Jabal Says. *Semitica et Classica* 2: 223.
al-Munajjid S.
1972. *Dirāsāt fī taʾrīkh al-khaṭṭ al-ᶜarabī*. Beirut: Dār al-Jīl.
Nehmé L.
(this volume). A glimpse of the development of the Nabataean script into Arabic based on old and new epigraphic material. In M.C.A. Macdonald (ed.), *The development of Arabic as a written language*. (Supplement to the Proceedings of the Seminar for Arabian Studies 40). Oxford: Archaeopress.
al-Qurṭubī, Abū ᶜAbd Allāh Muḥammad b. Aḥmad/[Editor unknown]
1952. *Al-jāmiᶜ li-aḥkām al-qurʾān wa-ʾl-mubayyin li-mā taḍammana min al-sunnah wa-āyāt al-furqān*. Cairo. [Publisher unknown].
al-Rāshid S.ᶜA.
1995. *Kitābāt islāmiyyah min makkah al-mukarramah: dirāsah wa-taḥqīq*. Riyadh: Maktabat al-malik fahd al-waṭaniyyah.
Robin C.J.
2006. La réforme de l'écriture arabe à l'époque du califat médinois. *Mélanges de l'Université Saint-Joseph* 59: 319–364.
Sharafaddin A.H.
1977. Some Islamic inscriptions discovered on the Darb Zubayda. *Atlal* 1: 69–70 [English], 73–74 [Arabic], pls 49–50.
al-Sijistānī, ᶜAbd Allāh b. Sulaymān b. al-ᶜAshᶜath, Abū Bakr Ibn Abī Dāwūd/[Editor unknown]
1936–. *Kitāb al-maṣāḥif*. Cairo. [Publisher unknown].
al-Ṣūlī, Abū Bakr Muḥammad b. Yaḥyā/ed. M.B. al-Atharī
1922. *Adab al-kuttāb*. Cairo: [Publisher unknown].
al-Suyūṭī, Abū ʾl-Faḍl ᶜAbd al-Raḥmān b. Abī Bakr/[Editor unknown]
1976. *Itqān fī ᶜulūm al-qurʾān*. Cairo. [Publisher unknown].
al-Ṭabarānī, Abū ʾl-Qāsim Sulaymān b. Aḥmad/[Editor unknown]
1984. *Al-muᶜjam al-kabīr*. Mosul. [Publisher unknown].
al-Ṭabarī, Abū Jaᶜfar Muḥammad b. Jarīr/[Editor unknown]
1967. *Jāmiᶜ al-bayān fī tafsīr al-qurʾān*. Cairo. [Publisher unknown].
Ṭarafah ibn al-ᶜAbd, ᶜAmr b. al-ᶜAbd b. Sufyān/ed. D. al-Khaṭīb & L. al-Saqqāl
1975. *Diwān*. Damascus: Majmaᶜ al-lughah al-ᶜarabiyyah.

Author's address
Professor ᶜAlī Ibrahīm Al-Ghabbān, The Saudi Commission for Tourism and Antiquities, PO Box 66680, Riyadh 11586, Kingdom of Saudi Arabia.

e-mail ghabbana@scta.gov.sa or ali.alghabban@scta.gov.sa

M.C.A. Macdonald (ed.), *The development of Arabic as a written language.* (Supplement to the Proceedings of the Seminar for Arabian Studies 40). Oxford: Archaeopress, 2010, pp. 103–112.

In search of a standard: dialect variation and New Arabic features in the oldest Arabic written documents

Pierre Larcher

Summary

The few surviving pre-Islamic inscriptions in both the Arabic language and the Arabic script show the absence of a standard written language. I will take as a sample of variation the inscriptions of Jabal Usays (AD 528–529) and Ḥarrān (AD 568). A meticulous examination of the first inscription suggests that its author, probably a soldier, wrote the way he spoke in a caseless variety of Arabic. In this context, I will examine the famous bilingual Greek-Arabic papyrus PERF 558 (22/643), in which the name *Ibn Abū Qīr* occurs twice. The Arabic *Abū Qīr* is the Greek *Apa Kyros*. In order to get to the form *Abū Qīr*, one has to go through the form *Abā Qīr, reinterpreted as the accusative of the three-case (triptotic) inflection Abū/Abā/Abī. Both types of Arabic, the so-called Old Arabic (inflected) and the so-called Neo-Arabic (non-inflected) coexisted, but the scribe used the New Arabic type. However, it is the Old Arabic type that was codified and became Classical Arabic. I will try to reach an understanding of the reasons for this choice.

Keywords: Arabic dialects, Old Arabic, *ta marbutah*, Jabal Usays, PERF 558.

Introduction

Arabic was spoken long before it came to be written, and for a long time it was written using various West Semitic and South Semitic alphabets. Then, in the sixth century AD, we find the first inscriptions which are in both the Arabic language and the Arabic script, and it is to these that the linguist must turn when trying to address the subject of this Special Session: "The development of Arabic as a written language".

1. The inscription of Jabal Usays[1]

Among the tiny amount of epigraphic evidence that has survived, there is a real gem, the inscription of Jabal Usays (Fig. 1).

It was originally published simply in facsimile, without a photograph (Fig. 2), by Muḥammad Abū ʾl-Faraj al-ʿUshsh in the journal *al-Abḥāth* (Beirut) (al-ʿUshsh 1964: 302–303, no. 85).

The first step in our understanding of this inscription took place in 1971, when it was republished by Adolf Grohmann with a facsimile and a photograph (Fig. 3) (Grohmann 1971: 15–17 and pl. 1/2).

Grohmann read the fourth line which gives the date in Nabataean numerals as 423 (4 × 100 + 20 + 3), which, if one takes the era to be that of the *Provincia Arabia* founded in March AD 106, represents AD 528–529. He therefore identified the king al-Ḥārith with the Ghassanid al-Ḥārith ibn Jabalah, who in 528 defeated the Lakhmid king al-Mundhir III. This makes this the oldest precisely dated inscription in both the Arabic language and Arabic script. The inscription of Zebed might claim this title, were it not simply an undated Arabic addendum to a Graeco-Syriac text of AD 512.

The second, and decisive, step was the rereading proposed by Christian Robin and Maria Gorea (Robin & Gorea 2002: 503–510), which I shall be using here (Fig. 4).

For the historian, the most exciting result of this rereading is the identification, at the beginning of line 3, of "Usays", the name of the place where the inscription was found. This ended the speculation as to whether the previous reading of this word, *sl(y)mn* ("Sulaymān/Salmān") was a personal name or a toponym (Grohmann 1971: 16; Shahid 1995: 117–124; MacAdam 1996: 49–57).

However, for the linguist, the most important result is the recognition of the syntactic structure "topic/comment" (in French "thème/propos", and the *mubtadaʾ*/

[1] This section returns to and develops the remarks on this inscription first made in Larcher 2005: 249–251.

Figure 1. *A photograph of the Jabal Usays graffito (courtesy of M.C.A. Macdonald).*

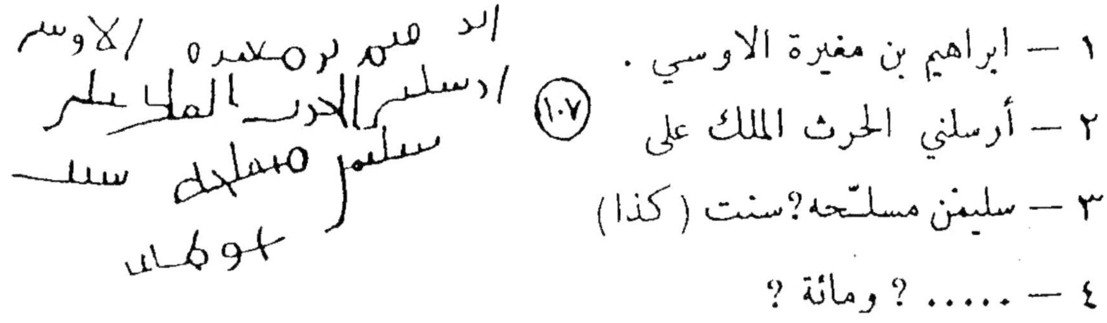

FIGURE 2. *The facsimile and transcription of the Jabal Usays graffito published by Muḥammad Abū ʾl-Farağ al-ʿUshsh (1964: 302, no. 85).*

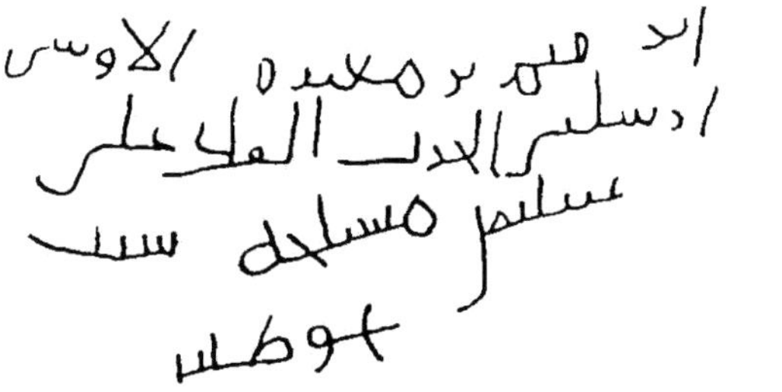

FIGURE 3. *The facsimile of the Jabal Usays graffito published by Grohmann (1971: 16, Abb. 7/d).*

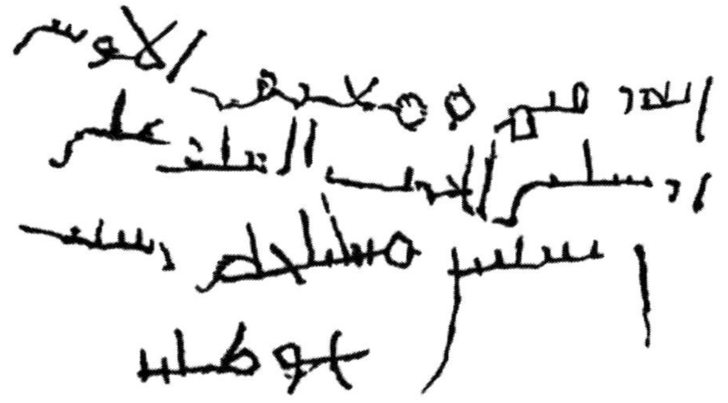

FIGURE 4. *The facsimile of the Jabal Usays graffito by Maria Gorea (published in Robin & Gorea 2002: 506).*

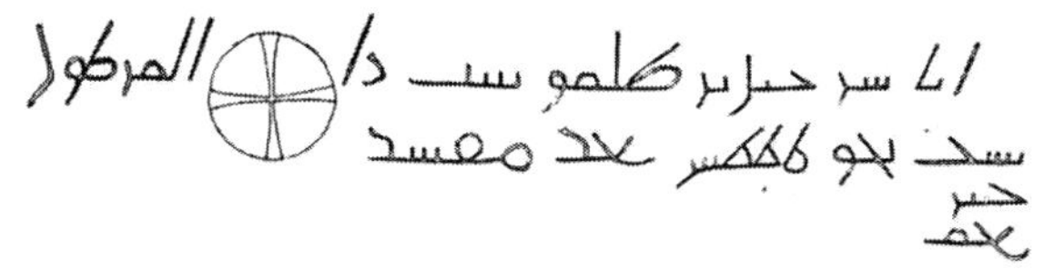

FIGURE 5. *The facsimile of the Ḥarrān inscription by Schroeder (1884: pl. 1).*

ḫabar of the Arab grammatical tradition): "I so-and-so, I did such-and-such". Thus, the Jabal Usays graffito has the same syntactic structure as the Ḥarrān inscription (AD 568) (Fig. 5), but in each the topic has a different form.

In the Ḥarrān inscription, the sequence *ʾalif*, *nūn*, *ʾalif* produces without difficulty the reading *ʾanā*, "I". In that of Jabal Usays, an *ʾalif* and a *nūn* without a point can be identified, but the third letter has to be a *hāʾ*, hence the reading *ʾanah* proposed by Robin and Gorea, which, as they note, could be a spelling of *ana*/*anā* attested in Aramaic and even in Arabic (2002: 508). Specialists in Arabic linguistics, familiar with the Arabic grammatical tradition, will immediately recall that the Arab grammarians regard *ʾanā* and *ʾanah* as the two *pausal* forms of the first person singular independent personal pronoun (*ḍamīr munfaṣil*). Thus, al-Zamakhsharī (died 538/1144) in his *Mufaṣṣal* writes: *wa-taqūlu fī ʾl-waqf ʿalā ghayr al-mutamakkina ʾanā bi-ʾl-ʾalif wa-ʾanah bi-ʾl-hāʾ* "With indeclinable words in pause we say *ʾanā* with *ʾalif* and *ʾanah* with *hāʾ*" (al-Zamakhsharī 1323/1905: 343). Raḍī al-dīn al-Astarābādhī (died 688/1289) in his *Sharḥ al-shāfiya* is more precise when he says, *wa-baʿḍ al-Ṭayyiʾ yaqif ʿalayhi bi-ʾl-hāʾ makān al-alif fa-yaqūlu ʾanah*, "Certain members of the tribe of Ṭayyiʾ mark the pausal form by substituting *hāʾ* for *ʾalif*" (al-Astarābādhī 1939–1958, ii: 294).

The attribution of this feature to the Ṭayyiʾ is apposite in the present context since the author of the Jabal Usays inscription calls himself *al-Awsī* (final word in line 1), i.e. a member of the tribe of Aws. Daniele Mascitelli shows that according to the Arabic sources there were three groups called Aws, and three others called al-Aws. Two, Aws b. Ḥārithah and al-Aws b. Dirmā were Qaḥṭānites and tribal sections of Ṭayyiʾ (Mascitelli 2006: 182). It is well known that the Syriac sources of this period and in this region used the term *Ṭayayē* as a metonym for the

Arabs, thus showing the importance of the tribe in this place and time. In any case, at this stage we can come to the initial conclusion that in the sixth century, there was no standard written language. The same morpheme could be expressed in two different forms, of which, at a later date, one became the norm in the classical morphology, phonology, and orthography.

We have just seen that *ʾanā* and *ʾanah* denote pausal forms. This is not at all surprising. The syntactic structure "topic/comment" constitutes what the Swiss linguist Charles Bally (1865–1947) called a segmented phrase in which the speaker pauses between the "topic" and the "comment" (Bally 1965: 61–62); and the presence of pausal forms confirms the identification of this structure here. But, at the same time, we can draw a second conclusion. Pausal forms belong above all to the spoken language, and the fact that they appear in writing suggests that we have identified a trace of the spoken in a written text. In this respect, it should be noted that in several cases, the type of Arabic called "Classical" tends to generalize *pausal* forms as *written* forms. As well as the first person pronoun which is everywhere written *ʾanā* (even though it is "normally" ʾana in context), another well-known case is that of the *tanwīnan*, always written as *-ā*, even though this spelling would "normally" represent only the pronunciation in pause.

The recognition of the pausal forms of the first person singular personal pronoun allows us to interpret more precisely another variant in the Jabal Usays inscription. At the end of line 3, the word *snt*, with a *tāʾ maftūḥah*, reflects the pronunciation of this word in all stages and varieties of Arabic, when it is the first term in an annexation (sanat kadhā). On the other hand, in line 1 where the author gives his name, Robin and Gorea (2002: 507–508) and Hoyland (2008: 55) read the patronym as Mughīrah with a *hāʾ*, which would correspond to the pausal pronunciation of the same suffix in Classical Arabic orthoepy even if today it has disappeared (*-ah* is realised simply as [-a]).[2]

This *-h* is also found in the second word of line 3, *mslḥh*, at least if this is to be read as *maslaḥah*. This reading is particularly appropriate in the geographical and historical context of the inscription; see Ibn Manẓūr (d. 711/1311) *Lisān al-ʿArab s.v.* S-L-Ḥ: *ka-ʾl-thaghr wa-ʾl-marqab* ("so to speak a frontier and an observation post") and, by metonymy, the soldiers (*qawm*) stationed there. However, it is less satisfactory on the syntactic level since it is difficult to see the relationship of *mslḥh* with the verb[3]. One solution would be to read *mslḥ-h*, i.e. *maslaḥ* with the third person singular masculine pronominal suffix, and translate "(Usays), his observation post". But *Lisān al-ʿArab* gives *maslaḥ* only as a toponym (*ism mawḍiʿ*), though it is possible that here *Lisān* is being normative. Nouns of place derived from nouns (denominatives, Arabic *ism al-kathrah, nomen abundantiae*) "normally" have the form *mafʿalah*, but modern usage provides plenty of exceptions to this "rule", e.g. *matḥaf* "museum", *maqhā* "café", etc.[4]

There is one other possibility. Lane (1863–1893) gives, on another authority (*al-Mughrib* of al-Muṭarrizī), a secondary meaning: "...and *maslaḥah* [also] is thus applied to a single person in a saying of Omar...". In *al-Mughrib* (a dictionary of legal terms) by al-Muṭarrizī (1328/1910: 259), under the lemma S-L-Ḥ there are various examples of the use of *maslaḥah*. Among them, the author quotes the second Caliph, ʿUmar ibn al-Khaṭṭāb, as follows: *wa-qawlu ʿumar raḍiya ʾllāhu ʿanhu: khayru ʾl-nās rajulun faʿala kadhā wa-kāna maslaḥatan bayna ʾl-muslimīna wa- ʾaʿdāʾihim* "in a saying of ʿUmar, may God be pleased with him: 'The best of people is a man who does such-and-such and acts as a frontier guard between the Muslims and their enemies'." If *maslaḥah* can refer to an individual, the sentence would simply and literally mean: "I, (name), sent me the king al-Ḥārith to Usays *as a frontier guard* in the year...." with *maslaḥah*

---

[2] However, M.C.A. Macdonald (personal communication and 2009*a*) points out that the patronym cannot be Mughīrah, since there is no medial *yāʾ* and the final letter has a long horizontal tail, making it almost certainly a *fāʾ*. He also remarks that the reading of the first name as Qutham (Robin & Gorea 2002: 507–508) or Qayyim (Hoyland 2008: 55) ignores the first letter which has to be *r/z*, a letter identical to the third letter of the patronym and the second letter in the word between the two names (hence *br*, not *bn* as suggested by Robin) (see Fig 1). He would therefore read the name as *rqym br mʿ/ghrf*, thus Raqīm/Ruqaym son of Muʿarrif. The fact that Muʿarrif occurs as a name in modern times was suggested by me to Macdonald in a message of 21 July 2009, and he points out that Robert Hoyland (who now also agrees with the reading) has found it in early Arabic texts, though so far not in the pre-Islamic era. Robin recently changed his translation to "Moi, Ruqaym fils de Muʿriḍ" (2008: 178). However, Macdonald argues that the reading of the final letter as a *ṣ/ḍ* is not possible (see 2009*a* and Fig. 1).

[3] Stefan Sperl (personal communication) suggests that it could result from the spontaneity of the writer. He wrote *Usays*, focusing on this word, before he added *maslaḥah* ("Usays, frontier post"). Had he thought before writing, he would have written *maslaḥat Usays* "the frontier post of Usays".

[4] Of course, it would be necessary to find some examples of *mafʿal* as a *nomen abundantiae* in the lexicon of the pre-Islamic era or, at least, of the early Islamic one. On the other hand if, following Grohmann (1971: 15, n. 3), we consider that there are four (not three) teeth after the *m* and therefore a supplementary letter, the word could be read either as *msylḥh* and interpreted as a diminutive of *mslḥh*, or as *mtslḥh* and interpreted as an attribute of the object *-nī*, i.e. as "his man at arms".

being read as the attribute of the object in a double-object construction "he sent me... as a..." normal in all forms of Arabic with many verbs.[5]

A third conclusion, at least if we accept the reading *Mughīrah* and/or *maslaḥah*, is that *tāʾ marbūṭah* did not yet exist. *Tāʾ marbūṭah* is a hybrid grapheme reflecting the double pronunciation of the suffix. One could only postulate its existence — allowing for the presence or absence of diacritical points — if the suffix were written *-h* (i.e. *snh*) throughout. Here, the author of the inscription wrote as *he pronounced*: *-t* in construct, but *-h* in pause.

Finally, when we combine the existence of the syntactic structure topic/comment and that of the pausal forms, we are led to a fourth conclusion. In the topic/comment structure, the topic has to be reiterated in the comment using an anaphoric pronoun (when the topic is a lexically full syntagme), or by a co-referent (when the topic is an independent pronoun of the first or second person, i.e. deictics). This is the case here, as in the Ḥarrān inscription. Here, the comment is a verbal clause, V[erb] O[bject] S[ubject], in which the topic is reiterated by the enclitic pronoun *-nī*. This immediately marks it as the object of the verb *ʾrsl* and consequently shows that the syntagme *ʾlmlk ʾlḥrth* is the subject of the verb. In other words, in such a structure, even if there had been a case ending, it would have been redundant; and the fact that it is redundant is a good reason for supposing that it was not present. In this context, the presence of pausal forms provides an additional argument. The spelling *Mghrh* (if we accept this reading) suggests that the author of the inscription was marking a pause, in other words that he was not linking the first appositional element of the topic (i.e. the syntactic group *qth/ym bn Mghrh*) to the syntactic group *ʾl-ʾwsy*, which follows it and which constitutes a second appositional element simply juxtaposed to the first. By contrast, in the most elevated register of Arabic, known as "Classical", it is the vowel of the case ending which simultaneously serves as the link between the two groups, thus: *ibnu Mughīrati l-Awsī*... Similarly, if we accept the reading *mslḥh* (*maslaḥah*), the spelling indicates that it marks a pause at the end of this phrase and before the one that follows. This would render Robin and Gorea's vocalization, *maslaḥatan*, slightly "surreal" (2002: 507).

Thus, a provisional conclusion on the Jabal Usays inscription would seem to be that the man who wrote it, wrote as he spoke and that he spoke a non-inflected variety of Arabic.

2. The Graeco-Arabic papyrus of 22/643.[6]

In this context, we now turn to the second series of the earliest surviving documents and in particular, to one of the two papyri dated to 22/643, the famous PERF (Papyri of the Archduke Rainer) 558 (= G. 39.726), in Vienna (Fig. 6).

This was published by Adolf Grohmann in 1932 in his "Aperçu de Papyrologie arabe" (1932: 40–43, and facsimile on plate 9), where he gives a transcription and translation.[7] It was recently republished by Demiri and Römer (2009: 8 for the translation, 9 for the Greek and Arabic texts, and 10 for the photograph of the recto) (Fig. 7).

This papyrus has attracted the attention of scholars for various reasons. Historians value it as contemporary evidence for the conquest of Egypt and the relations between the conquering and the conquered populations; while specialists in the history of the Arabic script have noted the presence of diacritical points and the dating by the Islamic era (Jones 1998: 95–103).

It consists of two versions, one in Greek and the other in Arabic, of an acknowledgement of delivery of meat to the troops of a Muslim emir by the authorities in Ahnās (Heracleopolis). The Arabic part is drawn up by a certain Ibn Ḥadīdū, with a *wāw* suffix which is also found much earlier in Nabataean inscriptions (see Cantineau 1934–1935: 77–97) and in the Ḥarrān inscription too (see Fig. 5), but which cannot be interpreted as a case ending. For it to be that, there would have to be at least one other such ending. For example, in Classical Arabic, the *w* of *muslimūn* is understood not only as a plural, but also as a nominative, because it is contrasted with *muslimīn*, where the *ī* marks simultaneously the plural number and the accusative/genitive case. On the other hand, in the Arabic dialects, where only 'muslimīn' occurs, the -īn represents only the plural, in contrast to the singular 'muslim.' If the *wāw* suffix is a very archaic feature, there is an innovation in comparison to the Jabal Usays inscription: the presence of *-h* in *khlfh* and *snh* in the construct state (*khalifat Tidraq/Uṣṭufur*, *sanat ithnayn wa-ʿishrīn*) shows that the *tāʾ marbūṭah* exists.

But the most remarkable thing is that the scribe, twice, writes *Ibn Abū Qīr*. This is not an "error", as it would be if, for example, the scribe had written one time Ibn

[5] I would like to express my thanks to the anonymous reviewer of this paper, who gave me this very interesting and useful reference, and for his (or her) comments.

[6] This section returns to and develops the remarks in Larcher 2007: 130.

[7] A photograph of the recto can be found in Ragheb 1996: 13.

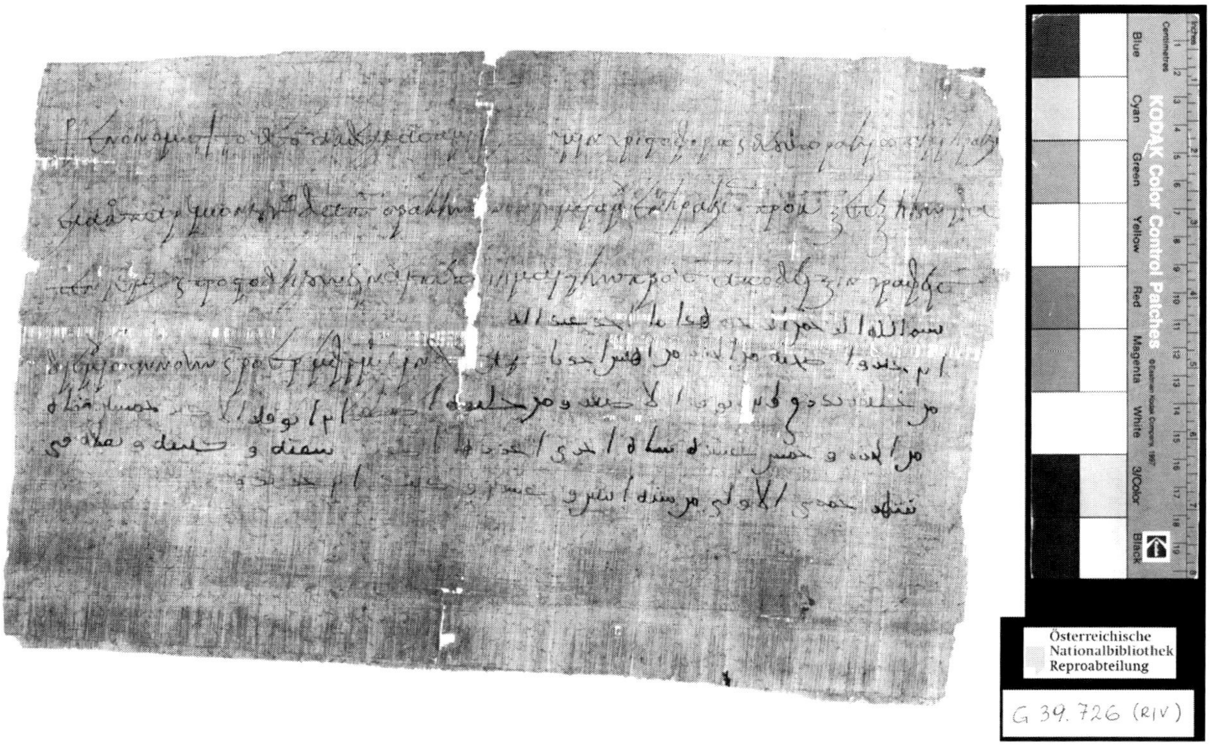

FIGURE 6. *PERF 558 recto (from http://www.onb.ac.at/sammlungen/papyrus/papyrus_bestandsrecherche.htm#).*

1. † ἐν ὀνόματι τοῦ θεοῦ ᾿Αβδέλλας ἀμιρ(ᾶς). ὑμῖν Χριστοφόρῳ καὶ Θεοδωρακίῳ παγάρχ(οις) Ἡρακλέ(ους).
2. ῎Ελαβο(ν) παρ᾿ ὑμῶν λόγῳ δαπ(ανημάτων) Σαρακηνῶν ὄντω(ν) μετά μου ἐν τ(ῇ) ʻἩρακλέ(ους) πρόβ(ατα) ξε, ἑξήκοντα
3. πέντε μό(να), (καὶ) πρὸς τὸ δῆλον εἶναι πεποίημαι τὴν παροῦσ(αν) ἀπόδειξιν γραφεῖσ(αν)

بسم الله الرحمن الرحيم هذا ما اخذ عبد اله

5. Δι᾿ ἐμοῦ ᾿Ιωάννου ν(ο)τ(αρίου) (καὶ) ροε (=διακόνου ?) † μ(ηνὸς) Φαρμ(οῦ)θ(ι) λ ἰνδ(ικτίο(νος) α †

ابن جبر و اصحبه من الجزر من اهنس اخذنا

من خليفة تذرق ابن ابو قير الاصغر ومن خليفة اصطفر ابن ابو قير الاكبر خمسين شاة

من الجزر وخمس عشرة شاة اخرى اجزرها اصحب سفنه وكتيبه وثقلاه في

شهر جمدى الاولى من سنة اثنين وعشرين وكتب ابن حديدو

FIGURE 7. *The transcription of the Greek and Arabic texts of PERF 558 (after Demiri & Römer 2009: 9).*

Abī Qīr and another time Ibn Abū Qīr. One could explain this "error" either in historical terms, as the appearance of a (uninflected) neo-Arabic type next to an (inflected) Old Arabic type, or in socio-linguistic terms, as the co-existence of two types (diglossia), i.e. in the first case, to regard it as a manifestation of Middle Arabic and, in the second, as a sign of Mixed Arabic.

However, *Abū Qīr* is not an Arabic name, but an Arabicized form of the Greek name: 'Απὰ Κύρος (*Apa Kyros*).[8] To get to *Abū Qīr*, one would have to pass through *Abā Qīr, which must then have been interpreted as the accusative of a triptotic inflection Abū/Abā/Abī (see also Diem 1984: 271–272 and n. 40). If the two types co-exist, it is in an asymmetrical way: there is an *implicit* reference to the Old Arabic type, but an *explicit* use of the Neo-Arabic type.

Before concluding, I would like to make one more observation. PERF 558 provides evidence not only for the absence of the visible case inflection (i.e. that shown by long vowels), but also the absence of the *invisible* case inflection (i.e. that represented by short vowels). In the Arabic, we have *ᶜbd ᵓlh*, with a single *l* — Grohmann (1932: 41) adds "*sic*" — but in the Greek we find "Abdellas"('Αβδέλλας), which, with its two lambdas, shows that Greek ears clearly heard two of them. Moreover, not only does the Greek write in one word what in Arabic is written as two, but the Greek also places the vowel e between the two elements. This is sufficient to demonstrate that the two parts were *understood* as what the French linguist André Martinet (1908–1999) calls a "syntheme" (Martinet 1960: 134), which was then treated as a radical to which the nominative case ending -ας (reserved for words of foreign origin) was added, and not as a syntagm, where both elements would be inflected. *ᶜAbd Allāh* being the subject of the verb, one should have a *-u-* between *ᶜAbd* and *Allāh*, and this would have been perfectly audible to Greek ears and could easily be transcribed in Greek as *ou.*[9]

Conclusions and hypotheses

I shall repeat here the main conclusions I have drawn so far:

1. The comparison of the Jabal Usays inscription with that of Ḥarrān shows that they both have the same syntactic structure — topic/comment — and the same topic, the first person singular personal pronoun. This, however, appears in the two pausal forms known from the Arabic grammatical tradition: *ᵓanah* in the Jabal Usays inscription, but *ᵓanā* in the Ḥarrān one. In the sixth century AD there was, then, no standard written language. The standardization begins when the form *ᵓanā* becomes the "Classical" form, in pause as well as in linking.

2. The syntactic structure topic/comment, which makes redundant — and therefore useless — a final case inflection, as well as the multiplication of pausal forms in the Jabal Usays inscription, suggest that the man who wrote this inscription wrote as he spoke, and that he spoke an uninflected variety of Arabic.

3. An examination of the Greek-Arabic papyrus of 22/643 (the oldest extant), confirms this interpretation. The Arab scribe of this papyrus twice uses *Ibn Abū Qīr* (and not *Ibn Abī Qīr).

4. *Abū Qīr* is not an Arabic name, but an Arabicized form of the Greek name *Apa Kyros*. To get to *Abū Qīr*, it would have had to pass through *Abā Qīr, which must then have been interpreted as the accusative of a triptotic inflection Abū/Abā/Abī. If one uses the traditional terminology of historical linguistics, though making it a purely typological interpretation, one may say that there is a co-existence of the two types, Old Arabic (inflected) and Neo-Arabic (uninflected), but with a difference in status: there is an implicit reference to the Old Arabic type, but explicit use of the Neo-Arabic type.

I shall add two hypotheses:

1. This asymmetry suggests that if in the seventh century AD, the two types co-existed, it would not be a case here of diglossia in the classic sense of the term, i.e. the co-existence of two varieties with complementary distribution of usage, as set out by the American linguist Charles Ferguson (1921–1998) (1959: 325–340). Rather, we have in the foreground, the Neo-Arabic type, an everyday vehicle of oral *and* written communication (which is why we find the dialect variation and the trace of the oral in the written); and, in the background, the Old Arabic type, which will become Classical Arabic. If, as one may assume, the latter was used solely for poetry, the case and modal inflection which is characteristic of it and which was probably inherited from a very distant past, was not syntactical, but prosodic.[10]

---

[8] On 'Απὰ/*Apa* see Derda & Wipszycka 1994: 23–56.

[9] There is another document, undated, but from the early Islamic, if not the pre-Islamic, era (see Macdonald 2009*b*: 50 and addenda), which shows the lack of case and mood endings: the famous bilingual fragment of a psalm from Damascus, published by Violet (1901), e.g. line 6 ΙΕΚΔΙΡ = *yaqdir* (not *yaqdiru*).

[10] The hypothesis that it was an innovation also has its partisans, for instance Owens (2006).

2. There remains Quranic Arabic: the Quranic *rasm* (the only objective fact for the linguist) shares with the archaic script its hyper-defectiveness (neither vocalization nor diacritical marks). This *rasm* displays non-classical features. The most well known, in the phonological sphere, is the *takhfīf al-hamzah*. This did not stop the grammarian readers (*qurrāʾ*) from "reading" *hamzah*, even where it is not marked and/or the rhyme shows that it was not "realised" (*taḥqīq al-hamzah*). Thus, for example, the reading of the written form *šīn* (without dots)–*yāʾ* (without dots)–*alif*, as *šayʾan* in sura 19: 9, 42, 60, 67, etc. The linguist of the twenty-first century will adopt an updated version of the hypothesis put forward by Karl Vollers (1857–1909) and emended by Paul Kahle (1875–1964), i.e. that the *qirāʾāt* represent the "classicization" of the Quranic language (Vollers 1906; Kahle 1947 [1959: 141–149]; 1948: 163–182; 1949: 65–71).

The form of Arabic called "Classical" is thus no different from all other classical languages: it is a *construct*.

Acknowledgement

Special thanks to M.C.A. Macdonald for allowing me to reproduce his unpublished photograph (Fig. 1).

Siglum

PERF von Karabacek 1894.

References

Al-Astarābādhī, Raḍī ʾl-dīn Muḥammad b. al-Ḥasan/ed. M.N. al-Ḥasan, M. al-Zafzāf & M.M. ʿAbd al-Ḥamīd
1939–1958. *Sharḥ shāfiyat Ibn al-Ḥājib* (4 volumes). Cairo. [Reprinted Beirut: Dār al-kutub al-ʿilmiyyah, 1395/1975].

Bally C.
1965. *Linguistique générale et linguistique française.* (Quatrième édition revue et corrigée). Berne: Francke.

Cantineau J.
1934–1935. Nabatéen et Arabe. *Annales de l'Institut d'Études Orientales* 1: 77–97.

Demiri L. & Römer C.
2009. *Texts from the Early Islamic Period of Egypt. Muslims and Christians at the First Encounter. Arabic Papyri from the Erzherzog Rainer Collection Austrian National Library Vienna* (Nilus, Studien zur Kultur Ägyptens und des Vorderen Orients, 15). Vienna: Phoibos.

Derda T. & Wipszycka E.
1994. L'emploi des titres *Abba*, *Apa* et *Papas* dans l'Égypte byzantine. *Journal of Juristic Papyrology* 24: 23–56.

Diem W.
1984. Philologisches zu den arabischen Aphrodito-Papyri. *Der Islam* 61: 251–275.

Ferguson C.
1959. Diglossia. *Word* 15: 325–340.

Grohmann A.
1932. Aperçu de papyrologie arabe. *Études de Papyrologie* 1: 23–95.
1971. *Arabische Paläographie.* ii: *Das Schriftwesen. Die Lapidarschrift.* Vienna: Böhlaus.

Hoyland R.
2008. Epigraphy and the Linguistic Background to the Qurʾān. Pages 51–69 in G.S. Reynolds (ed.), *The Qurʾān in its Historical Context.* London/New York: Routledge.

Ibn Manẓūr/ed. Y. Khayyāt
[n.d.]. *Lisān al-ʿArab al-muḥīṭ* (4 volumes). Beirut: Dār Lisān al-ʿArab.

Jones A.
1998. The Dotting of a Script and the dating of an era: The strange neglect of PERF 558. *Islamic Culture* 72: 95–103.

Kahle P.

1947. *The Cairo Geniza.* (Second edition 1959). Oxford: Blackwell.

1948. The Qurʾān and the *ʿArabīya.* Pages 163–182 in S. Löwinger & J. Somogyi (eds), *Ignace Goldziher Memorial volume.* Part I. Budapest.

1949. The Arabic readers of the Koran. *Journal of Near Eastern Studies* 8: 65–71.

Karabacek J. von

1894. *Papyrus Erzherzog Rainer: Führer durch die Ausstellung* (Arabische Abteilung). Vienna: Selbstverlag der Sammlung.

Lane E.W.

1863–1893. *An Arabic English Lexicon.* London: Williams & Norgate.

Larcher P.

2005. Arabe préislamique — Arabe coranique — Arabe classique: un continuum? Pages 248–265 in K-H. Ohlig & G-R. Puin (eds), *Die dunklen Anfänge. Neue Forschungen zur Enstehung und frühen Geschichte des Islam.* Berlin: Schiler.

2007. Les origines de la grammaire arabe, selon la tradition: description, interprétation, discussion. Pages 113–134 in E. Ditters & H. Motzki (eds), *Approaches to Arabic Linguistics Presented to Kees Versteegh on the Occasion of his Sixtieth Birthday.* (Studies in Semitic Languages and Linguistics, 49). Leiden: Brill.

MacAdam H.I.

1996. A note on the Usays (Jebal Says) Inscription. *Al-Abḥāth* 44: 49–57.

Macdonald M.C.A.

2009*a*. A note on new readings in line 1 of the Old Arabic graffito at Jabal Says. *Semitica et Classica* 2: 223.

Macdonald M.C.A.

2009*b*. *Literacy and identity in pre-Islamic Arabia.* (Variorum Collect Studies, 906). Farham. Ashgate.

Martinet A.

1960. *Éléments de linguistique générale.* Paris: Colin.

Mascitelli M.

2006. *L'arabo in epoca preislamica: formazione di una lingua* (Arabia Antiqua, 4). Rome: "L'Erma" di Bretschneider.

al-Muṭarrizī, Burhān al-Dīn Abū ʾl-Fatḥ Nāṣir b. Abī ʾl-Makārim ʿAbd ʾl-Sayyid. [Editor unknown].

1328/1910. *Al-Mughrib fī tartīb al-muʿrab.* Hyderabad: Maṭbaʿat Majlis Dāʾirat al-Maʿārif al-Niẓāmiyyah.

Owens J.

2006. *A Linguistic History of Arabic.* Oxford: Oxford University Press.

Ragheb Y.

1996. Les plus anciens papyrus arabes. *Annales Islamologiques* 30: 1–19.

Robin C.J.

2008. Les Arabes de Himyar, des "Romains" et des Perses (IIIe–VIe siècles de l'ère chrétienne). *Semitica et Classica* 1: 167–202.

Robin C.J. & Gorea M.

2002. Un réexamen de l'inscription arabe préislamique du Jabal Usays (528–529 è. Chr). *Arabica* 49: 505–510.

Schroeder P.

1884. Epigraphisches aus Syrien. *Zeitschrift der Deutschen Morgenländischen Gesellschaft* 38: 530–534, pls 1–2.

Shahid I.
1995. *Byzantium and the Arabs in the Sixth Century.* i/1. *Political and Military History*. Washington, DC: Dumbarton Oaks Research Library and Collection.
al-ᶜUshsh M.A.
1964. Kitābāt ᶜarabiyyah ghayr manshūra fī Jabal Usays. *Al-Abḥāth* 17: 227–316.
Violet B.
1901. Ein zweisprachiges Psalmfragment aus Damascus. *Orientalistische Litteratur-Zeitung* 10: 384–403; 11: 425–441; 12: 476–488.
Vollers K.
1906. *Volksprache und Schriftsprache im alten Arabien. Philologische Untersuchungen zur klassischen arabischen Sprache mit besonderer Berücksichtigung der Reime und der Sprache des Qorâns mit sieben Wörterverzeichnissen.* Strassburg: Trübner. [Reprinted Amsterdam: APA-Oriental Press, 1981].
al-Zamakhsharī, Abū ᵓl-Qāsim Maḥmūd b. ᶜUmar. [Editor unknown].
1323/1905. *Al-Mufaṣṣal fī ᶜilm al-ᶜarabiyyah wa-bi-dhaylihi al-Mufaḍḍal fī sharḥ ᵓabyāt al-Mufaṣṣal li-l-sayyid Muḥammad Badr al-dīn al-Naᶜsānī al-Ḥalabī.* Cairo. [Reprinted Beirut: Dār al-Ǧīl, n.d.]

Author's address
Pierre Larcher, Département d'études moyen-orientales, Université de Provence, 29, avenue Robert Schuman, F-13621 Aix-en-Provence Cedex, France.

e-mail Pierre.Larcher@univ-provence.fr

M.C.A. Macdonald (ed.), *The development of Arabic as a written language.* (Supplement to the Proceedings of the Seminar for Arabian Studies 40). Oxford: Archaeopress, 2010, pp. 113–120.

The codex Parisino-petropolitanus and the *ḥijāzī* scripts

François Déroche

Summary

The codex Parisino-petropolitanus, a Quranic manuscript from the third quarter of the first/seventh century, is the result of teamwork. The differences between the scripts of the five copyists who were involved in the transcription point to a lack of standardization in the matter of scripts during the first decades of caliphal rule. Although material evidence indicates that the manuscript is likely to have been produced for public use, each of the copyists had a personal approach to the transcription of the text, showing that this manuscript predates the emergence of specific, standardized Quranic scripts. This is further corroborated by each scribe's individual treatment of the orthography and the manner in which diacritical dots are used.

Key words: Arabic manuscripts, Arabic script, hijazi, diacritical dots, scribes

Teamwork seems to have been a rather common procedure in the copying of Quranic manuscripts during the early period (al-Aᶜẓamī 2003: 105; Rabb 2006: 84–127; Déroche 2009: 127–130). It is unfortunately not always easy to identify this situation when the fragments of a given manuscript — and so far we only know of fragmentary copies — do not contain a folio or a bifolio with two different scripts on it. With the codex Parisino-petropolitanus, we are fortunate enough to have an example of the hands of not merely two, but five scribes who co-operated in the transcription of this copy.

A few words about its history (Déroche 2009: 7–19): it was kept in the ᶜAmr mosque in Fusṭāṭ until, at the end of the eighteenth century, a member of the French expedition to Egypt, Jean-Joseph Marcel, acquired a few leaves which, in 1864, became part of the collection of the National Library of Russia, as it is known today. A few years later, another Frenchman, Jean-Louis Asselin de Cherville who served as a consular agent in Cairo from 1806 to 1822, was able to buy a larger number of folios, and these passed to the French National Library in 1833. Most of the ninety-eight folios can be found in those two collections: seventy in Paris[1] and twenty-six in St Petersburg.[2] Two more folios are known: one is in the Vatican library,[3] the other in the Khalili collection of Islamic art.[4] It is not known how they reached their present locations and it may be that other folios are still in private hands.

The manuscript in its present state contains slightly less than half the Quranic text; but even though two sequences of continuous quires are preserved, we are dealing with a series of fragments with occasionally conspicuous gaps.[5] Fortunately, four of the copyists worked almost simultaneously, one of them transcribing for instance the recto of a folio, then letting a colleague continue the text on the verso[6]. The co-operation of these four hands was therefore beyond any doubt. More problematic was the fifth scribe: his share of the common task constitutes the last part of what survives and is separated from the rest by a textual gap. The characteristics of his transcription — as we shall see later — were so different from the rest that, in the past, I came to the conclusion that it was part of another copy and described it separately in the catalogue of the Bibliothèque nationale (Déroche 1983: 59–60). It must be said that the folios of the Paris fragments were bound in such a way that it was almost impossible to undertake a very careful examination of the structure of the folios. The "discovery" of the sections of the manuscript in the National Library of Russia led me to revise my views:

[1] BNF Arabe 328 a and b (fols 1–70).

[2] NLR Marcel 18, fols 1–24 and 45–46.

[3] BAV Vat. Arab. 1605 [1]

[4] KFQ 60.

[5] The text runs almost continuously from sura 2: 275 to 3: 43; 3: 84 to 5: 33; 6: 20 to 10: 78; 10: 102 to 11: 35; and 12: 84 to 15: 87. Then comes a wider lacuna followed by some scattered sequences of folios: 23: 15 to 28: 53; 30: 58 to 31: 23; 35: 13–41; 38: 66 to 39: 55; 41: 31 to 46: 6; 56: 53 to 57: 26; 60: 7 to 63: 9; 65: 3 to 67: 26; and 69: 3 to 72: 2.

[6] See for example fols 9 r. and v. in BNF Arabe 328 (see Déroche & Noja Noseda 1998).

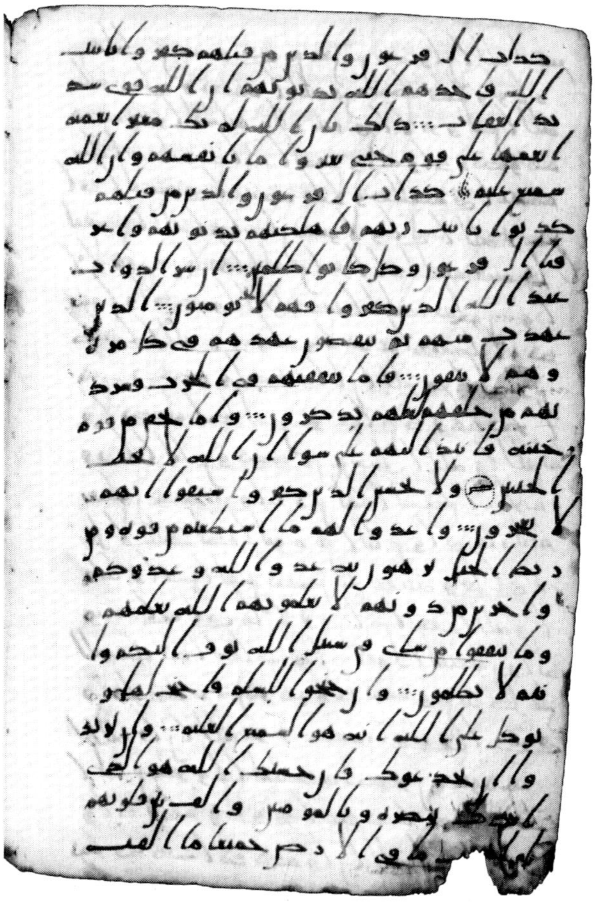

Figure 1. *MS St Petersburg, NLR, Marcel 18, fol. 2 v. (Qurʾān 8: 52–63). Hand A.*

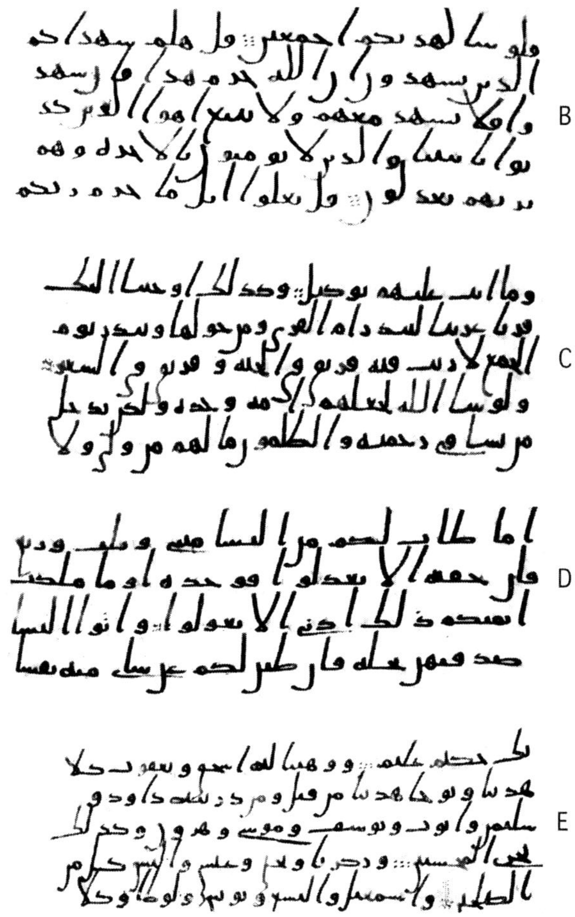

Figure 2. *Hands B, C, D, and E (after MS Paris, BNF Arabe 328).*

the fragments kept there are unbound, which allowed for an examination of the back of the quires and of the traces of the original sewing. It became obvious that the St Petersburg fragments belonging to the first part of the manuscript and the contribution of the fifth copyist were once part of one and the same volume. Actually, its material construction is quite coherent from beginning to end, with eight-folio quires in which the faces of the parchment are alternating (Déroche 2009: 22–26).

The codex Parisino-petropolitanus (Fig. 1) is a rather large manuscript: its folios measure about 330 by 250 mm. We can estimate that it had originally something like 210 or 220 folios, which means that between 17 and 18 m^2 of parchment would have been needed to produce it. This is to be compared with the 2.5 m^2 of a tenth-century Quranic manuscript from Qayrawān (Musée des arts islamiques, R 164) or the 3.5 m^2 of a copy produced in Malaga at the beginning of the fourteenth century (Escorial, MS 1397). The physical appearance of the codex Parisino-petropolitanus is an important element for a proper understanding of its meaning. It was a costly copy, even if the script seems today to be rather crude.

When analysing the individual contributions of the copyists (which I shall designate by letters from A to E; see Figs 1 and 2), we have first to take into account the fact that their respective sizes vary considerably. We cannot be sure that the present situation actually corresponds to the original division of labour since the whereabouts of half of the manuscript are unknown; or, at least, my attempts to find other folios have been unsuccessful. It may therefore well be that scribes D and E, who are each represented by a single folio, actually transcribed a larger amount of text; but we can only judge them from what has been preserved, that is to say very little. With this caveat, we can now turn to a presentation of the scripts.

At first glance, the difference between the five hands is

striking. It involves the copyist's personal way of writing, but it is also partly a matter of the writing tool used. This is particularly clear when it comes to measuring the height of the line in the various contributions and comparing it with the ratio between the width of the *calamus* tip and the height of the *alif*: D writes lines ranging from 14 to 15 mm, but his *alif* measures six times the width of his *calamus*, whereas E with smaller lines (10 to 13 mm) writes an *alif* which is nine times the dimension of the tip of his writing tool. On the other hand, some characteristics are shared by all. This is notably the case with the arrangement of the words on the page. Following the tradition of *scriptio continua* adapted to the Arabic script, the copyists do not try to differentiate the spaces between the letters of a word (when it contains a letter which is never connected to the next one) from those found between the words (Diem 1983: 386–387). They seem almost always to leave the same distance. In what could be a consequence of this way of handling the script, a word may be divided at the end of a line when there is not suffient space remaining to accommodate a word with one or more letters which are not joined to the left. The copyist will then write a first segment at the end of the line and the rest on the next one. It seems nevertheless that words are not divided into two parts at the end of a page. Ruling has been used for the preparation of the writing surface, but the number of lines to the page is far from homogeneous, ranging from 21 to 28. With 21 and 22 lines per page, B is certainly the copyist with the most amply spaced pages.

The quality of the script itself varies considerably from one copyist to the next. E is certainly the least gifted (Fig. 2): his lines seem to run downwards in the upper part of the page and then upwards in its lower part; the tops of the ascenders (a word I use cautiously here when dealing with the Arabic alphabet) often come into contact with the line of text above, and the left hand margin is very untidy. C and D, and particularly the latter, are professionals (Fig. 2). There is an obvious regularity in the script, a crispness that indicates a clear proficiency in this work. As regards the page, the result is balanced. A and B reflect an intermediary situation (Figs 1 and 2). They are not experts, as can be seen from the sometimes hesitant edge of the ascenders, but they try to write regularly.

The ascenders typically lean to the right, at an angle to the vertical varying in most cases from 20° to 35°. A few letters show variations from one hand to the other. *Alif* does not allow much variation since its shape is very plain. But the description by Ibn al-Nadīm, on which the definition of the *ḥijāzī* style is based, conferred a special status on this letter.[7] The height of the letter varies from one copyist to another, as we have seen above; the lower hook is less perceptible in B and D's contributions, for instance; and examples of ʿcurved final *alif*s are found in A. The inclination of the letter varies from 15° in the case of C to 35° with B.

Final and isolated *ṭāʾ* is written by A and B with a comma-like device protruding from the left lower end of the letter. The other scribes omit this element altogether. All write the ascender as leaning to the right, but B and C draw it slightly curved.

Qāf is somewhat difficult to handle when final or isolated, since its descending tail may interfere with the line below, and copyists like A and D try to control it as much as possible.

The isolated or final *kāf* is consistently written by all copyists in the same way: on the base line, a long stroke, then a sharp curve at its right extremity, connected to an upper stroke which is at a sharp angle to the lower one in A, B and E, but almost parallel to it in C and D; following a turn at 90°, the letter terminates with an ascender bent to the right, which C seems to draw slightly curved, as he does for *ṭāʾ*.

The final or isolated *mīm* comes in two varieties: one is circular, part of it below the line, and comes with a tail written almost on the line (a shape found in B and E); the other is also circular, but it has no tail or only a tiny one (mainly in D). There are intermediary shapes in A and C.

The final *nūn* — and this also applies to *sīn/shīn* and *ṣād/ḍād* — is written by A, D, and sometimes also by C, almost like a "J", while the other copyists tend to draw a half-circle.

As early as 1856, Michele Amari could write that the letter forms of the codex Parisino-petropolitanus and a papyrus dated 40 of the hijrah (Silvestre de Sacy 1833: 66–68)[8] were "bearing as great a resemblance to each other as is possible between a handwritten private letter and large copies of a sacred book intended for mosques" (Amari 1910: 16). More recently, Adolf Grohmann stated that, "this style of writing [i.e. *ḥijāzī*] is ... a secular script" (1958: 221–222). A look at the papyrus of Ahnās (ÖNB PERF no. 558) shows some significant similarities, although the writing surface is quite different from that of parchment and had some influence on the ductus (Larcher,

[7] Ibn al-Nadīm 1970: 10; 1971: 9. For the discussion of the text and definition see Abbott 1939: 17–18; and recently Déroche 2009: 108–117.

[8] This document, which Amari examined personally, seems to have been lost.

this volume: Fig. 6; Robin 2006: fig. 7, for instance). In addition to the slanting of the ascenders, which is slighter in the papyrus than in the manuscript, we notice that the connection between the letters is treated in a similar way by the scribe of the Ahnās papyrus and by B. The shape of the letters *dāl*, *qāf*, and *nūn* are significantly close in both the papyrus and the manuscript. We have to turn to other documents in order to examine the isolated or final *kāf*, which is usually written with a wider opening of the angle but keeps the distinctive extended stroke on the base line (e.g. Khan 1992: 36, ill. for *c*.135 and 140 AH).

The relative closeness between the script of the documents and that of the copyists of the codex is certainly true for A or B, but not so much for D. Although similarities do exist between the script of D and that of his colleagues, his contribution, limited as it is, stands alone in the codex. The writing instrument used by D has a broader nib and the control over the script — notably the clear spacing between the lines — is obvious. Like D, C is a professional, but the ascenders he draws are not formed so consistently: their orientation is far from homogeneous. We have to turn to other Quranic fragments like BNF Arabe 330 c (Déroche 1983: 144–145; 2004: 117–118 and fig. 42) or NLR Marcel 13 (Déroche 2006) in order to find parallels, as far as the script is concerned. When considering the shape of the letters found in the two pages by D, one notices that many of them — final *mīm* and *nūn* are particularly clear examples — are close to their regularized forms found in Marcel 13. The exception is the *hāʾ*, which has a clearly different shape.

D's contribution is a key element for the dating of the codex Parisino-petropolitanus. It is obviously earlier than Marcel 13 and the fragments related to it: the script has not yet reached the same level of quality, but this would not be sufficient evidence in itself as the codex might have been transcribed in a remote place, far from the latest fashion in the matter of script. Other characteristics should also be taken into account: the orthography is on the whole slightly less developed in D than in Marcel 13 (Déroche 2006: 238–240 and table 2). In the codex Parisino-petropolitanus, there is no illumination between the suras (Déroche 2009: 30), which is no longer the case in Marcel 13 and related fragments (Déroche 2004: 116, 117–118 and fig. 42; 2006: 242–253 and figs 3–9). In addition, some of the sura-headings found there include elements that are part of an Umayyad decorative repertory. This set of characteristics suggests that Marcel 13 and related fragments are slightly later than the codex and might reflect the impact of ʿAbd al-Malik's reforms. On the other hand, the Parisino-petropolitanus is obviously a copy of an earlier original (Déroche 2009: 153)[9] and its orthography is the result of an improvement of that found in the model (2009: 152–153). Thus it cannot belong to the earliest stage of the textual transmission. For these reasons, it can tentatively be attributed to the third quarter of the first century of the hijrah.

A striking feature of the codex is the way in which the copyists dealt with the changes of hand. In all cases in the extant fragments, the copyists finished their task at the bottom of a recto. This means that there is no instance of an opening with two different hands facing each other. This is quite probably a consequence of the awareness of the differences between the various styles and the possible imbalance which would strike the reader if, when opening the copy, he would have had before him the verso of a folio in one style and the opposite page in another

The use of the diacritical marks is of special interest. A quick comparison between a folio of the manuscript and the same portion of the text in a modern edition shows a huge gap in this respect: on fol. 44 r., for instance, A chose to identify eight of the 240 dotted letters of the printed version. The question is now: are there rules in the use of the diacritical marks? Were some letters singled out and if so, why? The answer is somewhat complicated by the fact that we are dealing with five copyists who did not share the same view about this. Their individual approach can be summarized in the following table (Fig. 1).

Letters bearing diacriticals	**Copyists**				
	A	**B**	**C**	**D**	**E**
Bāʾ		x			
Tāʾ	x	x		x	
Thāʾ	x	x	x	x	
Jīm					
Khāʾ	x				
Dhāl	x	x	x	x	
Zay	x	x		x	
Shīn	x	x			
Ḍād	x				
Ẓāʾ	x				
Ghayn	x				
Fāʾ		x			
Qāf					
Nūn	x	x	x	x	
Yāʾ		x			

[9] Some scribal mistakes can only arise from the transcription of an exemplar.

It is immediately obvious that E did not use any diacritical points at all on his two pages, and that C — the second best represented hand in terms of volume of text transcribed — only distinguishes three letters, namely *thāʾ*, *dhāl*, and *nūn*, and this occurs only five times in sixteen folios. We also notice that two letters which should normally be marked, *jīm* and *qāf*, are never indicated by any of the four hands using diacritical marks.[10] With *qāf*, the distinctive final form was perhaps an element weighing against systematic pointing, although within a word the same grapheme is used for both *fāʾ* and *qāf*. Only B dotted *fāʾ*, and we know from other fragments of that period that one of the solutions for distinguishing these letters was to point one of them and leave the other unpointed. Turning to the other copyists, the table also makes clear that they do not point the same letters. Leaving aside D, since we only have two pages by him, the comparison between A and B shows that the former never points below the line, and thus *bāʾ* and *yāʾ* are never distinguished in the folios he transcribed.

Turning now to the frequency of the use of diacritical marks for the individual letters, one seems to have received more attention than any other: *nūn* is by far the most frequently pointed but, strangely enough to a modern observer, not primarily in places which could be ambiguous for the reader. Thus *-nā*, either as the ending of the first person plural of the perfect tense of the verb or as a possessive suffix, is identified by a dot in 233 cases, and to these could be added the instances of words with a final *nūn*, like *muʾmin*, with the undetermined accusative case ending (*tanwīn*), which are graphically similar to the *-nā* ending mentioned above. To this we can also add the medial *nūn* in *ʿinda* (eighteen times), in *min-hu* or *-hā*, *-kum*, and *-hum* (thirty-two cases) and *inna-hu* (twenty-five). Similarly, but less frequently, *dhāl* in *ʾidh* and *ʾidhā* is marked (forty-nine cases).

We would nowadays assume that the initial grapheme of a verb in the imperfect tense would have been the object of much attention and that the copyists would have tried to dispel any doubt, particularly concerning the second and third persons of the singular and plural. Even in the case of A who does not put dots below the line, we could expect that a significant number of verbs beginning with *tāʾ* or *nūn* would have been pointed. However, this is clearly not the case. Among the three copyists who used diacritical marks to any significant extent, i.e. A, B, and D, I can find only eleven instances of verbs in the imperfect or related forms with one or two dots providing a clue as to the correct reading of the initial letter. One of the aims attributed to the Uthmanic recension was to avoid divergent readings of the text: in the codex Parisino-petropolitanus, the use of the diacritical marks remains negligible and considerably below what would have been necessary to eliminate any ambiguity.

On the other hand, one cannot say that the copyists systematically avoided putting marks on possibly disputed graphemes in order to leave open the reading of the text. In 3: 180, we find *taḥsabanna* instead of *yaḥsabanna*,[11] a canonical *qirāʾa* that differs from the Ḥafṣ *ʿan* ʿĀṣim text. The other instances prove unsurprising. Even in the case of a verse, which is known to have been involved in early controversies like 10: 22, the dots found there were put on a verb which was not part of the debate.[12] Among the eleven cases of diacritical pointing on the initial letter of a verb in the imperfect, the third person is never pointed, even though four of these cases occur in parts of the text due to B, who does sometimes point *yāʾ*. In these cases, B, like A, added dots to the *tāʾ* of the second person or *nūn*, but not to *yāʾ*. In this small sample, the relatively high frequency of verbs in the energetic mood (in 3: 180 and 188 with the same verb; 7: 134; and 10: 22) should be noted, although the significance of this is uncertain. Anyhow, the copyists did sometimes take sides.

The distribution of the dotted letters in the codex does not seem to answer a specific purpose: one can find pages without any dot at all and conversely a line with a concentration of diacritical marks. On NLR Marcel 18, fol. 16 r., for instance, on lines 13 and 14 corresponding to sura 26: 50 and 51, there is an "accumulation" of dots, which mainly identify a *nūn* in unambiguous situations, like *rabbunā*. To characterize the use of diacriticals as random would perhaps be excessive, but it was clearly — as with the script — a matter of personal choice on the part of the copyists. Although one of them does not use any dots at all, it seems highly likely that they were a normal feature of the Arabic script at the time the codex Parisino-petropolitanus was written. This implies that any degree of central control was quite unlikely, although we are here dealing with Quranic codices, which should have been a central concern for the caliphate. Once again, it should be noted that the manuscript was not a private copy but a large *muṣḥaf* that was probably meant for some sort of public use.

As a result of the collaboration of five copyists, this

[10] A similar situation for *jīm* and *qāf* has been observed on inscriptions and documents from the same period (see Robin 2006: 344). On the other hand, *ḍād* and *ẓāʾ* are pointed in the manuscript.

[11] BNF Arabe 328, fol. 8 v.

[12] BNF Arabe 328, fol. 46 r.

manuscript provides us with a unique opportunity to make a contrastive study of the use of writing during this early period. If the dating suggested — the third quarter of the first century (roughly between AD 670 and 695) — is correct, it means that, until the early years of Umayyad rule, the script, including the use of diacritical marks, as well as the orthography were largely left to individual taste in the case of Quranic manuscripts, although various accounts show that public authorities were already involved in the production and diffusion of copies of the text. Each of the copyists of the codex Parisino-petropolitanus had a highly recognizable style, which suggests that, when learning how to write, they were never taught a common repertory (D being perhaps an exception) and that the changes of hand did not matter for those who paid for what was a relatively costly copy. The manuscript is witness to a period antedating the emergence of the concept of Quranic scripts and, more generally, of a controlled use of the Arabic script.

Sigla

BAV	Bibliotheca apostolica vaticana
BNF	Bibliothèque nationale de France
KFQ	The Nasser D. Khalili collection of Islamic art
NLR	National Library of Russia
ÖNB	Österreichische Nationalbibliothek
PERF	Papyrus Erzherzog Rainer collection
Vat. Arab.	Vaticanus Arabicus

References

Abbott N.

1939. *The rise of the North Arabic script and its ḳurʾānic development. With a full description of the Ḳurʾān manuscripts in the Oriental Institute.* (The University of Chicago Oriental Institute publications, 50). Chicago: University of Chicago Press.

Amari M.

1910. Bibliographie primitive du Coran…Extrait de son mémoire inédit sur la chronologie et l'ancienne bibliographie du Coran, publié et annoté par Hartwig Derenbourg. Pages 1–22 in *Centenario della nascita di Michele Amari* i. Palermo: Virzi.

al-Aʿẓamī M.M.

2003. *The history of the Qurʾānic text: from revelation to compilation. A comparative study with the Old and New Testaments.* Leicester: UK Islamic Academy.

Déroche F.

1983. *Les manuscrits du Coran: Aux origines de la calligraphie coranique.* (Bibliothèque nationale, Catalogue des manuscrits arabes, 2e partie, Manuscrits musulmans, 1/1). Paris: Bibliothèque nationale.

2004. *Le livre manuscrit arabe. Préludes à une histoire.* (Conférences Léopold Delisle). Paris: Bibliothèque nationale de France.

2006. Colonnes, vases et rinceaux. Sur quelques enluminures d'époque omeyyade. *Comptes rendus de l'Académie des inscriptions et belles-lettres* 2004: 227–264.

2009. *La transmission écrite du Coran dans les débuts de l'islam. Le codex Parisino-petropolitanus.* (Texts and Studies on the Qurʾân, 5). Leiden: Brill.

Déroche F. & Noja Noseda S.

1998. *Le manuscrit arabe 328 (a) de la Bibliothèque nationale de France.* (Sources de la transmission manuscrite du texte coranique, i: Les manuscrits de style higâzî). Lesa: Fondazione Ferni-Noja Noseda.

Diem W.

1983. Untersuchungen zur frühen Geschichte der arabischen Orthographie. IV. Die Schreibung der zusammenhängenden Rede. Zusammenfassung. *Orientalia* [NS] 52: 357–404.

Grohmann A.

1958. The problem of dating early Qurʾâns. *Der Islam* 33: 213–231.

Ibn al-Nadīm/transl. B. Dodge.
1970. *The Fihrist of al-Nadīm. A tenth-century survey of Muslim culture.* i. New York: Columbia University Press.

Ibn al-Nadīm/ed. R. Tajaddūd.
1350/1971. *Kitāb al-fihrist.* Tehran: [Publisher unknown].

Khan G.
1992. *Selected Arabic papyri.* (Studies in the Khalili collection, 1). London: Nour Foundation/Oxford University Press.

Rabb I.
2006. Non-Canonical Readings of the Qur'ān: Recognition and Authenticity (The Ḥimṣī Reading). *Journal of Qur'ānic Studies* 8/2: 84–127.

Robin C.J.
2006. La réforme de l'écriture à l'époque du califat médinois. *Mélanges de l'Université Saint-Joseph* 59: 319–364.

Silvestre de Sacy A.I.
1833. Mémoire sur deux papyrus, écrits en langue arabe, appartenant à la collection du Roi. *Mémoires de l'Institut royal. Académie des inscriptions et belles-lettres* 10: 65–88.

Author's address
François Déroche, École pratique des hautes études, Sciences historiques et philologiques, 45, rue des Écoles, 75005 Paris, France.

e-mail francois.deroche@dbmail.com

M.C.A. Macdonald (ed.), *The development of Arabic as a written language.* (Supplement to the Proceedings of the Seminar for Arabian Studies 40). Oxford: Archaeopress, 2010, pp. 121–130.

The relationship of literacy and memory in the second/eighth century

Gregor Schoeler

Summary

The production, presentation, and transmission of knowledge in Islamic society of the second/eighth century was almost entirely oral, although notebooks were kept as aides-memoire by scholars, poets, and their transmitters. The only literary works published in writing and intended to be read in this century were those of the *kuttāb* (scribes or state secretaries) whose literary output consisted both of original literary works (*rasāʾil*, epistles), and adaptations of Middle Persian works. These writings were composed for a readership, albeit one that, in this century, consisted exclusively of the caliph, his court, and the circle of *kuttāb*. Traditional Muslim scholars (i.e. scholars learned in the fields of religion, philology, and history) published their materials not through composing books for a *readership*, but through "academic instruction" (audition, dictation, or student recitation), even in the case of works subdivided into thematic chapters (*muṣannafāt*). This "lecture system", which is a characteristic feature of Muslim scholarly culture, has been difficult for modern scholars to appreciate and this has led to an incorrect assessment by modern Western and Muslim scholars of the nature and authenticity of early Islamic tradition.

Key words: Literacy, Memory, early Islam, writing, aural transmission, orality

The relationship of literacy and memory (or orality) in early Islamic culture is complex and has been difficult for scholars to appreciate (Schoeler 2009: 2–9). We are confronted with the problem especially in the second/eighth century, when it comes to the production/creation, presentation, and transmission of knowledge, Arabic *ʿilm*, which also means "science" (Anon 1971). In what follows, I mean by "knowledge" both scholarly forms ("sciences") and literary forms, including poetry, keeping in mind that poetry was conceived of by traditional critics as *ʿilm al-ʿarab* ("the knowledge" or "science of the Arabs"), sometimes even as *akbar ʿulūm al-ʿarab* ("the greatest of the sciences of the Arabs") (Heinrichs 1969: 55–56).

I

Pure orality, in second/eighth century poetry (as at any time, for that matter), was in evidence when a poet improvised verses. Production and recitation coincided, with the poet having to produce, and at the same time recite, a poem, either spontaneously, without any prior thinking (*irtijālan*), or after short reflection (*badīhan*) (Schoeler 2006: 93–94). Since the poet was pressed for time, impromptu poems, called *qiṭaʿ* (sg. *qiṭʿah*), were almost always short (Schoeler 2006: 95, 101; 2004). When it came to long poems, called *qaṣāʾid* (sg. *qaṣīdah*), there was possibly still evidence of pure orality in tribal poetry, the production/creation, recitation, and transmission of which had been initially — and for a long time remained — purely oral. But Dhū ʾl-Rummah (d. 117/735, see Sezgin 1967–2007, ii: 394–397), a well-known Umayyad poet who was dubbed "the last Bedouin", is nevertheless said to have known how to read and write (al-Marzubānī 1965: 280; Schoeler 2006: 68). He is reported to have instructed three scholars and transmitters, *ruwāt* (sg. *rāwiyyah*) — among them Ḥammād al-Rāwiyah (d. *c.*156/773) — who took dictation from him or read his poems back to him so that he could correct the texts they had recorded. It is reported that Dhū ʾl-Rummah pointed out the mistakes in their notes. When the scholars expressed surprise that he knew how to write, Dhū ʾl-Rummah explained that a "settled" scribe had visited him in the desert and taught him by drawing letters in the sand.

We even know why the use of writing by Bedouin poets was frowned upon. The literary theorist al-Kalāʿī (fl. *c.*542/1148) says:

> They feared that he [sc. the poet] would be unnatural and affected by using the reed-pen and have recourse to his sense of sight for (poetic) speech, since (when a poet writes) those two [sc. reed-pen and sense of sight] are part of the work, and play a role in (the process of) composition. (1966: 235–236)

According to this point of view, writing is not needed as a support by someone endowed with natural poetic talent. Poets working with reed-pen and papyrus/paper were considered to be "unnatural", "affected" and were regarded by certain scholars as less talented than those who eschewed these tools (Schoeler 2006: 68–69).

It is quite clear that, at this time, the recitation of poetry, which at the same time constituted its "publication", was carried out exclusively orally by the poet or his transmitter (*rāwī*); but anecdotes such as the one I have just quoted, permit us to assume that written notes were in fact used to bolster the memory, at least in preserving and transmitting the poems, and that some poets perhaps also wrote when composing a poem, a process which was often slow and laborious.

What applied to tribal poetry held true, to an increasing degree, for the poetry of "settled" poets and court poetry. It is well known (Schoeler 2009: 21–22) that the great Umayyad poet al-Farazdaq (d. 110/728, see Sezgin 1967–2007, ii: 359–363) owned notebooks containing the poems he had learned in order to obtain *riwāyah*, i.e. knowledge of the poetry of his great predecessors so that, as a poet, he would be capable of transmitting it further (Heinrichs 1969: 49). In one of his famous *naqāʾiḍ* (poetic "flytings" or polemical poems), al-Farazdaq explicitly says, "I possess the written compilation of the poems of al-Ǧaᶜfarī [= Labīd] and the earlier Bishr [ibn Abī Ḥāzim]" (Bevan 1905–1912, i: 201, no. 39, verse 57, see also verse 61).

Al-Farazdaq's transmitter (*rāwī*), Mattawayhi, is reported to have written down the poems of his master (Bevan 1905–1912, ii: 903, l. 1). This indicates that written records were used to preserve and transmit the poems. Of the other great Umayyad poet, Jarīr (d. 111/729, see Sezgin 1967–2007, ii: 356–359), we learn that, when he wanted to compose a lampoon, he told his transmitter Ḥusayn, "Put more oil into the lamp today and prepare tablets and inks" (Bevan 1905–1912, i: 430, l. 12). This shows that Jarīr used to dictate the verses when he had composed them, apparently to bolster his memory.

In the second/eighth century, then, the poet or his *rāwī* always relied on his memory while *reciting* poems. However, when *preserving* poetry in order to have it at his disposal or to transmit it further, he used written records, or notebooks. Moreover, in the process of composing (long) poems, *qaṣāʾid*, some "settled" poets at least seem to have written down or dictated the verses they had produced/created. Neither the poets nor their transmitters found it necessary, in this century, to produce finalized editions of their poetry. Even in the third/ninth and fourth/tenth centuries most poets left it to later generations to collect their poems and to edit them in a *dīwān*, although they used written notes and even had written collections (Schoeler 2009: 3).

II

Pure literacy proper, and written transmission, were to be found in the second/eighth century in the works of the *kuttāb* (sg. *kātib*, scribes or state secretaries) and the representatives of the great wave of translation, first from Middle Persian, later from Syriac and Greek (Schoeler 2009: 56–58; Heinrichs 1969: 40–42). The *kuttāb* comprised a new social class, or scholarly cadre, of non-Arab descent, which, in that century, had appeared on the scene and taken its place beside the traditional Muslim scholars (i.e. scholars learned in the fields of religion, philology, and poetry) (Gibb 1962). These new Muslims, mostly of Iranian extraction, worked in the state administration, and had ideas and ideals completely different from those of the scholars engaged in religious and linguistic scholarship. Their literary output consisted both of original literary works and also translations, or rather, adaptations. Initially, their original literary works invariably took the form of epistles, *rasāʾil* (sg. *risālah*), and consequently were addressed to a specific person. This holds true for all the works of ᶜAbd al-Ḥamīd al-Kātib (d. 132/750, see Latham 1983: 164–169), for example, and of at least one work of Ibn al-Muqaffaᶜ (d. *c.* 139/757, see Latham 1990: 48–77), namely the *Risālah fī ʾl-ṣaḥābah*.

The scribes naturally composed their epistles — which were in literary, we might even say artistic, prose — in *writing*. Indeed, these letters (or books) were conceived of entirely with the prospect of written transmission in mind. Wadād al-Qāḍī has been able to show that although copies of ᶜAbd al-Ḥamīd's epistles already circulated in his own lifetime, they were disseminated to a much greater extent after his death, principally in the circles of the Abbasid secretaries, where they were used as a kind of textbook, or pedagogical model; they were also passed on by his numerous sons, one of whom became a famous secretary in his own right in early Abbasid times (al-Qāḍī 1992: 215–275).

These writings were composed for a readership, although it is true that, in the second/eighth century, this readership consisted exclusively of the caliph, his court, and the circle of the secretaries. It was not until the third/ninth century that writers, or rather men of letters, produced literary *rasāʾil* (as well as large books) for a wider lay reading public (Schoeler 2009: 99–107). The most famous of these was al-Jāḥiẓ (d. 255/868–869, see Pellat 1965).

III

And yet, notwithstanding the presence of the works by the secretaries, during the entire second/eighth century and before, the dissemination of knowledge was effected by traditional Muslim scholars (i.e. scholars learned in the fields of religion, philology, and poetry), not through actual books intended for a reading public, but through systematic teaching, "academic instruction" (Sezgin 1967–2007, i: 58–60; Schoeler 2006: 33–36, 40–42; 2009: 40–42, 69–71, 76). This instruction took place in two ways: through *samāᶜ*, or "audition", which involved the students listening to a teacher's recitation; or through a variant of audition termed *qirāʾah*, "recitation", which was a lecture during which a student recited material on a subject in the presence of a teacher, and the teacher listened and made corrections. Recitation amounted to scholarly transmission. This "lecture system" is a characteristic feature of Muslim scholarly culture and was maintained, with only some modifications, throughout the entire pre-modern period (Schoeler 2009: 122–125).

In the second/eighth century there was vehement discussion among Muslim scholars about whether or not it was permitted to use written records as lecture notes or mnemonic aids (Cook 1997; Schoeler 2006: 111–141). The discussion centred specifically on the issue of the permissibility of writing down hadiths, literally "sayings", reports, or "traditions" about the words or deeds of the Prophet Muḥammad or one of his companions. Many scholars took the view that prophetic hadiths in particular, but also other material, were only to be memorized and transmitted orally, their main argument being that the Qurʾān should be and should remain the only book in Islam (Schoeler 2006: 84–85, 116–117). In the first half of the second/eighth century, this opinion seems to have dominated the entire Islamic world; but in the second half of that century most non-Iraqi scholars gave up this view, although it remained dominant in the scholarly centres of Iraq (2006: 114–116, 126–127). Some explicitly advocated the position that hadiths and other reports could be put into writing without reservation (2006: 116, 129). But this concession referred to notes serving as aides-memoire, not to edited books. I cannot deal here with all aspects of this debate, but I want to discuss some far-reaching consequences of the (controversial) prohibition on writing down traditions.

In the middle of the second/eighth century *taṣnīf*, a new method of arranging and presenting knowledge, established itself (Sezgin 1967–2007, i: 55–58; Schoeler 2009: 68–84). *Taṣnīf* consisted in classifying traditions and other material according to subject matter: works so organized were called *muṣannaf* (pl. *muṣannafāt*). Their publication was effected in the traditional way, in lectures in scholarly circles, through audition and recitation (Schoeler 2009: 69–71, 76–77). Many scholars in the Iraqi centres of learning, Basra and Kufa in particular (Schoeler 2006: 114–116; Cook 1997: 444–459), continued to recite hadiths and other reports from memory, refusing to rely on notes or notebooks as aides-memoire. Of the Basran Saᶜīd b. ᶜArūbah (d. 156/773, see Sezgin 1967–2007, i: 91–92), who is reported to have been the first, or one of the first, compilers of a *muṣannaf* work (Ibn Ḥajar 1978, *Muqaddimah*: 5; Schoeler 2009: 68, 69), we learn that he "had no book, but kept everything in his memory" (Ibn Ḥajar 1984–1985, iv: 57; al-Dhahabī *c.*1963, ii: 153).

But could he realistically have memorized what must have been a very substantial collection of hadiths? The biographical literature (Ibn Saᶜd 1904–1940, vii b: 76) does tell us that Saᶜīd had a scribe named ᶜAbd al-Wahhāb b. ᶜAṭāʾ, who always accompanied him and who wrote down his "books". It appears likely therefore that, before teaching, Saᶜīd would consult and retrieve the material for his lecture from a certain number of writings; this material would not be taken from writings belonging to him, there being no such thing, but from those in his scribe's possession.

In Kufa, memorization of traditions was *de rigueur* until at least the first half of the third/ninth century (Schoeler 2006: 115; 2009: 70; Cook 1997: 454–459). Concerning the Kufan traditionist and compiler of a *muṣannaf* work, Wakīᶜ b. al-Jarrāḥ (d. 197/812, see Sezgin 1967–2007, i: 96–97) we are told, "Wakīᶜ ... was one of those who travelled [sc. in search of knowledge, i.e. hadiths], wrote down, collected, classified, memorized, recapitulated and disseminated" (Ibn Ḥibbān 1959: 173, no. 1374).

Thus Wakīᶜ, unlike Saᶜīd b. Abī ᶜArūbah, used to write down the hadiths he had "heard" from his shaykhs. Another account has it that he did this not during the lectures of his masters but afterwards, at home. He is reported to have said, "I never used to write down a hadith from Sufyān, but committed it to memory. Upon returning home, I wrote it down" (al-Khaṭīb al-Baghdādī 1931, xiii: 475).

Yet, he only used his collection to bolster his memory, since the "publication" and further transmission of his work had to be carried out through recitation from memory. This interpretation is confirmed by another statement concerning Wakīᶜ, "We never saw a book in Wakīᶜ's hands, since he would recite his books from memory" (Ibn Ḥibbān 1973–1983, vii: 562).

The fact that Wakīᶜ (and other traditionists too) did take notes while collecting traditions but never used them while lecturing shows that the use of written records was, at that time and place, considered permissible for private, not public, purposes. Some thought it worse to own written materials than to dictate such materials to others (Schoeler 2009: 48, 56). There was even a widely held view that it was permissible to write down traditions provided one then erased them (Schoeler 2006: 113; Cook 1997: 487–488). If one owned any writings, these had to be destroyed before one died, or at the latest, shortly thereafter (Schoeler 2006: 113).

Even in the third/ninth century, when the production of literary works had begun on a truly large scale, several Iraqi traditionists are said to have recited their teachings from memory (Schoeler 2006: 115; 2009: 70). For instance, the Kufan, Ibn Abī Shaybah (d. 235/849), states at the beginning of several chapters of his monumental *Muṣannaf* (1966–1983, x: 154; xiv: 388, 424, 427, 512, 539), "This is what I know by heart from the Prophet" (*hādhā mā aḥfaẓu min al-nabī*).

This odd way of expressing oneself shows that, even at a time in which the records of the traditionists had grown to manuscripts comprising many volumes, some Iraqi scholars still felt obliged to present their compilations as aides-memoire, i.e. as writings for purely private use.

In Medina, as well as in other centres, personal and oral (or, rather, aural) instruction remained the norm when it came to publishing and transmitting systematically classified works. This was accomplished by audition, dictation, or by student recitation (Schoeler 2009: 71–73). This is true, for instance, of *Kitāb al-Maghāzī* ("The Book of the Campaigns") of Ibn Isḥāq (d. 150/767, Sezgin 1967–2007, i: 288–290), the most important classical biography of the Prophet Muḥammad, and the *Muwaṭṭaʾ* ("The Well-Trodden [Path]", an eminent corpus of juridical material) by Mālik b. Anas (d. 179/796, see Sezgin 1967–2007, i: 457–465). Information on Ibn Isḥāq's teaching and transmission practices is relatively plentiful. One of his students, Yūnus ibn Bukayr (d. 199/815, see Sezgin 1967–2007, i: 146, no. 91), a transmitter who prepared a recension of his teacher's work, says:

> The whole of Ibn Isḥāq's narrative (*ḥadīth*) is "supported" (*musnad*) [i.e. it is based on Ibn Isḥāq himself], since he dictated it to me (*amlāhu*) or recited it in my presence [from a notebook?] (*qaraʾahū ᶜalayya*) or reported it to me [from memory?] (*ḥaddathanī bihī*). But what is not "supported" is recitation [by a student] (*qirāʾah*) in the presence of Ibn Isḥāq. (al-ᶜUṭāridī 1978: 23).

In contrast to the practice in Basra and Kufa, the opposition to the writing-down of traditions had disappeared after the middle of the second/eighth century in Medina and most of the other non-Iraqi scholarly centres (Schoeler 2009: 70–73; Cook 1997: 461–462). Unlike their Iraqi colleagues, these compilers of *muṣannaf* works did not feel the need to hide any written collections they had in their possession and even used them in public without hesitation. For instance, Maᶜmar b. Rāshid (d. 154/770, Sezgin 1967–2007, i: 290–291), a Basran who settled in Yemen, would "care for his books and consult them" since, in that part of the Muslim world, memorization of hadiths was not especially prized. Whenever he had occasion to return to his hometown of Basra, however, Maᶜmar found himself obliged to recite the hadiths from memory (Ibn Ḥajar 1984–1985, vi: 279).

To return to the transmission practices in Medina, Salamah b. Faḍl (d. 191/806), another student of Ibn Isḥāq, is said to have prepared a copy of the whole *Kitāb al-Maghāzī* for his teacher, which the latter then collated against his own copy (Ibn ᶜAdī 1988, iii: 140; Ibn Ḥajar 1984–1985, iv: 135). Moreover, Salamah is reported to have inherited all the manuscripts in Ibn Isḥāq's estate (al-Khaṭīb al-Baghdādī 1931, i: 220f.); as a result, he — and he alone — used his teacher's autograph copies in the subsequent transmission of the work. This explains why it is that Salamah — on whose authority, for instance, the famous universal historian and Qurʾān commentator al-Ṭabarī (d. 310/923, see Bosworth 2000; Sezgin 1967–2007, i: 323–328) quotes Ibn Isḥāq — is credited with having put together the "most complete books of the *Maghāzī*" (Ibn Ḥajar 1984–1985, i: 135; al-Dhahabī *c.*1963, ii: 92).

What remains of this and other *muṣannaf* works?[1] None of them has come down to us in its original form; at most, we have later recensions of these compilations, but none dating from earlier than the third/ninth century. All of them were transmitted and reworked by a student or, more commonly, by a student of a student of the compiler. In the best cases, these transmissions "on the authority of so-and-so" form the basis of independent works: for instance the *Sīrat Muḥammad rasūl Allāh* ("Biography of Muḥammad, the Apostle of God"), which is a recension of Ibn Isḥāq's *Kitāb al-Maghāzī* by Ibn Hishām (d.

[1] For the following, see Schoeler 2009: 76–79.

218/834, Sezgin 1967–2007, i: 297–299), or the so-called "Vulgata" of the *Muwaṭṭaʾ*, which is a recension of Mālik's work by Yaḥyā b. Yaḥyā al-Maṣmūdī (d. 234/848, Sezgin 1967–2007, i: 459–450). In other cases, these materials appear as quotations in later compilations; for instance al-Ṭabarī's numerous — often quite lengthy — quotations from Ibn Isḥāq in the chapters of his *Taʾrīkh al-rusul wa-ʾl-mulūk* ("History of the Apostles and Kings") dealing with the *Maghāzī*.

The nature of these compilations — the *muṣannafāt* (as well as related works) — has been extremely difficult for scholars to grasp fully (Schoeler 2009: 2–9, 76). Since none of them has survived in its original form, we are entitled to ask whether they are examples of literature properly speaking. We can answer this question by turning to Greek literature, which is possessed of works akin to these *muṣannafāt*. W.W. Jaeger has described these as "neither lecture notebooks, nor literature", "elaborated" writings to be sure, but "not ones intended for publication with a larger lay reading public in mind". They are examples of "a systematic literature of the school, for the school ... published... through lectures" (1912: 135–137, 145, 147). Jaeger's description of the Greek teaching texts could just as easily be applied to our *muṣannafāt*: they too are examples of a literature of the school, intended solely for use by the school, and published through recitation, i.e. through audition, dictation, or recitation by a student. Thus, these works were not yet "literature" but at a later stage of development, became literature. The transmission process did not consist of copying finalized books, although written notes were used; memory mattered in this process, since an author could "publish" his work by reciting it by heart and students might write down what they had heard, not immediately, but later on, from memory. Still, there is one *muṣannaf* work in the second/eighth century, certainly an exception, which is, in fact, an actual book. Indeed, it is the very first book, properly speaking, in all of Islamic scholarship, namely the *Kitāb* of Sībawayhi (d. *c.*180/798, see Carter 1997; Sezgin 1967–2007, ix: 51–63), a comprehensive and systematic description of Arabic grammar. Sībawayhi conceived of this work as a written text with a reading public in mind (Humbert 1997: 555; Schoeler 2009: 87–89). This is shown *inter alia* by the presence of cross-references in the text. The *Kitāb* was, it is true, disseminated through lectures whenever possible, like the other scholarly output of that time; however, thanks to the fact that it was a fixed text, its transmission did not depend any longer on oral (or rather aural) transmission. Muslim scholars were quick to recognize the uniqueness of the *Kitāb* as an actual book, and some even called it "the Qurʾān of grammar" (Abū ʾl-Ṭayyib 1955: 65).

To reiterate: the problem of the relationship of literacy and memory in the scholarly and written culture of the second/eighth century concerns above all the *muṣannaf* works compiled by traditional scholars in the second half of that century. To this, we can add the collections of poetry by the learned transmitters.[2] While the literary works — mostly epistles composed by state secretaries and translations from foreign languages — were intended as written works destined for a reading public and thus disseminated by copying finalized manuscripts, the *muṣannaf* works (books subdivided into thematic chapters), being compilations of traditions and other material, were "published" by systematic teaching. This was effected by scholars through lectures. Some scholars knew the lectures by heart, but possessed and used written notes as aides-memoire, others read out their notebooks publicly. The students took dictation or wrote down later what they had heard, typically at home. Only exceptionally do we learn that a student copied the manuscript of his master and used it in the further transmission of the work.

This system of instruction and transmission, which had originated in the last third of the first/seventh century — ʿUrwah b. al-Zubayr (d. *c.*94/712, see Schoeler 2000; 2009: 41–43) being one of the first and certainly the most important scholar to apply it — has been difficult for modern scholars to appreciate. This lack of understanding has largely been responsible for an incorrect assessment of the nature of early Islamic tradition by modern Western (and Muslim) scholars. A long-standing discussion concerning the oral or written nature of this tradition has arisen, especially among German-speaking scholars, starting with Ignaz Goldziher (1890). Given that "oral" was often identified, consciously or unconsciously, with "unreliable", the transmission of knowledge in early Islam was considered by most of these scholars to have been inauthentic. In any case, the two distinct (though related) issues of literacy versus orality and reliability versus unreliability were often intermingled. Reacting against the position of the advocates of the oral (or unreliable) nature of early Muslim tradition, Fuat Sezgin (1967–2007, i: 53–84) and Nabia Abbott (1957–1972) attempted to demonstrate that the transmission was basically written, and then inferred, from its putative written nature, its authenticity (to a large extent, at least).

[2] For instance, the *Muʿallaqāt*, allegedly compiled by Ḥammād al-Rāwiyah, and the *Mufaḍḍaliyyāt* compiled by al-Mufaḍḍal al-Ḍabbī.

Abbott's and Sezgin's view notwithstanding, in the Anglophone world it was essentially the position of Goldziher and Joseph Schacht (1950; 1953) which was followed. Goldziher had maintained that the Islamic tradition was only the result of the theological, social, and political tendencies of later times in which these materials (so he assumed) originated. Under Goldziher's influence, Schacht had argued that the corpus of the Islamic traditions, legal but also historical, originated no earlier than the second/eighth century and so could not contain reliable information on the first/seventh century, for instance on the life of Muḥammad. From then on, most Anglophone Islamicists assumed that the Islamic transmission system in its entirety was unreliable. One reason for this view, but certainly not the only one, was the putative orality of the transmission. Patricia Crone (1980: 3–17) inferred the unreliability of the entire early Islamic tradition, on the one hand, from the historical situation ("circumstances of drastic changes") and, on the other, from its oral nature (1980: 4). "The tradition is full of contradictions, confusions, inconsistencies, and anomalies" (1980: 12). Michael Cook maintained that there are no objective criteria for authenticity in the study of early Islamic literature (1981: 156; 1983: 61–63). John Wansbrough argued that traditional accounts of early Islamic history were nothing but salvation history (1977; 1978). There were dissenting voices, such as those of Montgomery Watt (1964; 1983) and Robert B. Serjeant (1990) for instance, but the "sceptics" won the day. This, in spite of the fact that Albrecht Noth, who was long reckoned by the sceptics to be one of their number, had argued that the compilations contain both "good" and "bad" traditions and that Muslim tradition "offers much material which, if in need of careful examination, is still of historical value" (1968: 295; 1994: 24).

What was disregarded, or omitted by the "sceptics", was an extensive and close study of the peculiarities of the early Islamic transmission system. Such research has been undertaken, in recent decades, by Harald Motzki (1991; 2002), Stefan Leder (1991), Sebastian Günther (1991), and others, including myself. As I hope I have demonstrated in this paper, the teaching and transmission of knowledge in early Islam was neither merely oral nor merely written. This is true of the second/eighth century lecture system I have described, but also of its beginnings in the last third of the first/seventh century. The nature of the transmission neither assured complete or far-reaching reliability, as Abbott and nowadays Sezgin assume, nor did it exclude it, as many Western sceptics maintain. The issues of literacy versus orality and reliability versus unreliability have to be treated separately. The challenge is to find criteria of authenticity — beyond orality versus literacy — in order to be able to distinguish "good" and "bad" traditions. I make an attempt to meet this challenge in my book *Character und Authentie der muslimischen Ueberlieferung über das Leben Mohammeds* (Schoeler 1996), which will appear in an English translation (forthcoming) soon, *in shāʾ Allāh.*

References

Primary Sources

Abū ʾl-Ṭayyib al-Lughawī, ᶜAbd al-Wāḥid b. ᶜAlī/ed. M.A. Ibrāhīm.
1955. *Marātib al-naḥwiyyīn.* Cairo: Maktabat Nahḍat Miṣr.

Bevan A.A. (ed.).
1905–1912. *The Naḳāʾiḍ of Jarīr and al-Farazdaq.* (3 volumes). Leiden: Brill.

al-Dhahabī, Shams al-Dīn Abū ᶜAbd Allāh Muḥammad b. Aḥmad/ed. ᶜA.M. al-Bijāwī.
*c.*1963. *Mīzān al-iᶜtidāl fī naqd al-rijāl.* (4 volumes). Beirut: Dār al-maᶜrifah.

Ibn Abī Shaybah, Abū Bakr ᶜAbd Allāh b. Muḥammad/ed. ᶜA. Khān al-Afghānī.
1966–1983. *Al-Kitāb al-muṣannaf.* (15 volumes). Bombay: al-Dār al-Salafiyyah.

Ibn ᶜAdī, Abū Aḥmad ᶜAbd Allāh/ed. S. Zakkār.
1988. *Al-kāmil fī ḍuᶜafāʾ al-rijāl.* (8 volumes). (3rd edition). Beirut: [publisher unknown].

Ibn Ḥajar al-ᶜAsqalānī, Shihāb al-Dīn Abū l-Faḍl Aḥmad b. ᶜAlī/ed. Ṭ.ᶜA. Saᶜd & M.M. al-Hawārī.
1978. *Fatḥ al-bārī bi-sharḥ ṣaḥīḥ al-Bukhārī.* (*Muqaddimah* and 28 volumes). Cairo: Maktabat al-kulliyyāah al-Azhariyyah.

Ibn Ḥajar al-ᶜAsqalānī, Shihāb al-Dīn Abū l-Faḍl Aḥmad b. ᶜAlī/ [editor unknown].
1984–1985. *Tahdhīb al-tahdhīb.* (14 volumes). Beirut: [publisher unknown].

Ibn Ḥibbān al-Bustī, Abū Ḥātim Muḥammad/ed. M. Fleischhammer.
1959. *Kitāb mashāhīr ᶜulamāʾ al-amṣār.* Cairo/Wiesbaden: Steiner.

Ibn Ḥibbān al-Bustī, Abū Ḥātim Muḥammad/ed. ᶜA. Khān.
1973–1983. *Kitāb al-thiqāt.* (9 volumes), Hyderabad: Dāʾirat al-maᶜārif al-ᶜuthmāniyyah.
Ibn Saᶜd, Abū ᶜAbd Allāh Muḥammad/ed. E. Sachau, C. Brockelmann, J. Horovitz, J. Lippert, *et al.*
1904–1940. *Kitāb al-ṭabaqāt al-kabīr. Biographien Muhammeds, seiner Gefährten und der späteren Träger des Islams bis zum Jahre 230.* (9 volumes). Leiden: Brill.
al-Kalāᶜī, Muḥammad b. ᶜAbd al-Ghafūr/ed. M.R. al-Dāyah.
1966. *Iḥkām ṣanᶜat al-kalām.* Beirut: [publisher unknown].
al-Khaṭīb al-Baghdādī, Abū Bakr Aḥmad b. ᶜAlī/[editor unknown].
1931. *Taʾrīkh Baghdād.* (14 volumes). Cairo: Maktabat al-Khanjī.
al-Marzubānī, Abū ᶜUbayd Allāh Muḥammad b. ᶜImrān/ed. ᶜA.M. al-Bijāwī.
1965. *Al-muwashshaḥ.* Cairo: Dār Nahḍat Miṣr.
al-ᶜUṭāridī, Abū ᶜUmar Aḥmad b. ᶜAbd al-Jabbār/ed. S. Zakkār.
1978. *As-Siyar* = (Ps.-)Muḥammad b. Isḥāq, *Kitāb al-siyar wa-ʾl-maghāzī.* Beirut: Dār al-fikr.

Secondary Sources

Abbott N.
1957–1972. *Studies in Arabic Literary Papyri.* (3 volumes). Chicago, IL: University of Chicago Press.
Anon.
1971. ᶜIlm. Pages 1133–1134 in *Encyclopaedia of Islam.* (New edition). iii. Leiden: Brill.
Bosworth C.E.
2000. Al-Ṭabarī. Pages 11–15 in *Encyclopaedia of Islam.* (New edition). x. Leiden: Brill.
Carter M.G.
1997. Sībawayhi. Pages 524–530 in *Encyclopaedia of Islam.* (New edition). ix. Leiden: Brill.
Cook M.
1981. *Early Muslim Dogma: a Source-Critical Study.* Cambridge: Cambridge University Press.
1983. *Muhammad.* Oxford: Oxford University Press.
1997. The Opponents of the Writing of Tradition in Early Islam. *Arabica* 44: 437–530.
Crone P.
1980. *Slaves on Horses. The Evolution of the Islamic Polity.* Cambridge: Cambridge University Press.
Gibb H.A.R.
1962. The Social Significance of the Shuubiyya. Pages 62–73 in H.A.R. Gibb, *Studies on the Civilization of Islam.* (Ed. S.J. Shaw & W.R. Polk). Boston, MA: Beacon.
Goldziher I.
1890. Ueber die Entwickelung des Ḥadīth. Pages 1–274 in I. Goldziher, *Muhammedanische Studien.* ii. Halle: Niemeyer. [Translated as: On the Development of the Ḥadīth. Pages 17–251 in I. Goldziher, *Muslim Studies.* ii. (Trans. S.M. Stern). Chicago, IL: Allen & Unwin, 1971].
Günther S.
1991. *Quellenuntersuchungen zu den "Maqātil aṭ-Ṭālibiyyīn" des Abū l-Farağ al-Iṣfahānī (gest. 356/967).* Hildesheim: Olms.
Heinrichs W.
1969. *Arabische Dichtung und griechische Poetik.* Beirut/Wiesbaden: Steiner.
Humbert G.
1997. Le Kitāb de Sībawayhi et l'autonomie de l'écrit. *Arabica* 44: 553–567.
Jaeger W.W.
1912. *Studien zur Entstehungsgeschichte der Metaphysik des Aristoteles.* Berlin: Weidmann.
Latham D.
1983. The Beginnings of Arabic Prose Literature: The epistolary genre. Pages 154–179 in A.F.L. Beeston, T.M. Johnstone, R.B. Serjeant & G.R. Smith (eds), *Arabic Literature to the End of the Umayyad Period.* (The Cambridge History of Arabic Literature). Cambridge: Cambridge University Press.

1990. Ibn al-Muqaffaᶜ and Early ᶜAbbāsid Prose. Pages 48–77 in J. Ashtiany, T.M. Johnstone, J.D. Latham, R.B. Serjeant & G.R. Smith (eds), *ᶜAbbāsid Belles-Lettres.* (The Cambridge History of Arabic Literature). Cambridge: Cambridge University Press.

Leder S.

1991. *Das Korpus al-Haitham ibn ᶜAdī (st. 207/822). Herkunft, Überlieferung, Gestalt früher Texte der Akhbār Literatur.* Frankurt am Main: Klostermann.

Motzki H.

1991. *Die Anfänge der islamischen Jurisprudenz. Ihre Entwicklung in Mekka bis zur Mitte des 2./8. Jahrhunderts.* (Abhandlungen für die Kunde des Morgenlandes, 50/2.) Stuttgart: Steiner.

2002. *The Origins of Islamic Jurisprudence: Meccan Fiqh before the Classical Schools.* (Trans. M. Katz). Leiden: Brill.

Noth A.

1968. Iṣfahān-Nihāwand. Eine quellenkritische Studie zur frühislamischen Historiographie. *Zeitschrift der Deutschen Morgenländischen Gesellschaft* 118: 274–296.

1994. *The Early Arabic Historical Tradition.* Princeton, NJ: Darwin.

Pellat C.

1965. Al-Djāḥiẓ. Pages 385–387 in *Encyclopaedia of Islam.* (New edition). ii. Leiden: Brill.

al-Qāḍī W.

1992. Early Islamic State Letters: The Question of Authenticity. Pages 215–275 in A. Cameron & L.I. Conrad (eds), *The Byzantine and Early Islamic Near East.* i. *Problems in the Literary Source Material.* Princeton, NJ: Darwin.

Schacht J.

1950. *The Origins of Muhammadan Jurisprudence.* Oxford: Clarendon.

1953. On Mūsā ibn ᶜUqba's Kitāb al-Maghāzī. *Acta Orientalia* 21: 288–300.

Schoeler G.

1996. *Charakter und Authentie der muslimischen Überlieferung über das Leben Mohammeds.* Berlin/New York: De Gruyter.

2000. ᶜUrwa b. al-Zubayr. Pages 910–913 in *Encyclopaedia of Islam.* (New edition). x. Leiden: Brill.

2004. Ḳiṭᶜa. Pages 538–540 in *Encyclopaedia of Islam.* (New edition). xii. (Supplement). Leiden: Brill.

2006. *The Oral and the Written in Early Islam.* (Trans. U. Vagelpohl, ed. J.E. Montgomery). London/New York: Routledge.

2009. *The Genesis of Literature in Islam. From the Aural to the Read.* (In collaboration with and translated by S. Toorawa). Edinburgh: Edinburgh University Press.

(forthcoming) *The Biography of Muḥammad. Nature and Authenticity.* Trans. U. Vagelpohl, ed. J.E. Montgomery. London/New York: Routledge.

Serjeant R.B.

1990. Meccan Trade and the Rise of the Islam: Misconceptions and Flawed Polemics. *Journal of the American Oriental Society* 110: 472–486.

Sezgin F.

1967–2007. *Geschichte des arabischen Schrifttums.* (13 volumes). Leiden: Brill.

Wansbrough J.

1977. *Quranic Studies. Sources and Methods of Scriptural Interpretation.* Oxford: Oxford University Press.

1978. *The Sectarian Milieu. Content and Composition of Islamic Salvation History.* Oxford: Oxford University Press.

Watt W.M.

1964. The Materials Used by Ibn Isḥāq. Pages 23–34 in B. Lewis & P.M. Holt (eds), *Historians of the Middle East.* Oxford: Oxford University Press.

1983. The Reliability of Ibn Isḥāq's Sources. Pages 31–43 in *La vie du prophète Mahomet.* Colloque de Strasbourg (octobre 1980). Paris: [publisher unknown].

Author's address

Emeritus Professor Gregor Schoeler, Orientalisches Seminar der Universitaet Basel, Maiengasse 51, CH-4056 Basel, Switzerland.

e-mail gregor.schoeler@unibas.ch

M.C.A. Macdonald (ed.), *The development of Arabic as a written language*. (Supplement to the Proceedings of the Seminar for Arabian Studies 40). Oxford: Archaeopress, 2010, pp. 131–140.

The use of the Arabic script in magic

Venetia Porter

Summary
This paper focuses on the Arabic script from the perspective of the form it takes when used in magic. Particularly characteristic is the use of early angular letter forms which are often written in continuous lines, whose meanings have either been variously interpreted or which have no obvious meaning at all. These inscriptions appear on a variety of magical objects, such as amulets and magic bowls. The amulets are inscribed either in positive or in negative, perhaps in order to be stamped onto something, although it is not clear what this might be. By being in reverse it may simply be that this enhances their magical obscurity. This paper will look in detail at this script and attempt to examine why this form was so popular in magic. It will also place it within the context of other "magical" scripts in use in the Islamic period.

Keywords: magic, Kufic, amulet, magical alphabets, mysterious letters

Among the vocabulary of magical elements, which includes five- or six-pointed stars, symbols, numbers, and strange words, is the use of an early Arabic script. Angular in style, this script appears in a form described by Casanova in an article on magical alphabets (1921) as "Linear Kufic" (*Koufique linéaire*). The shapes of the letters are typical of those found in inscriptions from the first two centuries of Islam and are generally combined to form letter strings. The letters themselves can range from being recognizable to highly schematized. The "magical vocabulary" as Savage-Smith has described it becomes fully established by the twelfth–thirteenth centuries AD, and is used on objects in a variety of contexts from magic bowls to architecture (Savage-Smith 1997: 60) and Linear Kufic plays an important role within it. Key questions, not all of which can be answered at the present time, are how and why the angular Kufic style in this form was adopted as a "magical" script, and how early; what types of objects does it appear on; to what extent can the inscriptions be deciphered; how the form of these magical inscriptions relates to non-magical Arabic inscriptions; and if and how this script relates to other magical alphabets in use in the early Islamic period.[1]

Letter magic

The first question, however, concerns letter magic. The reason for the popular use of letters for amulets lay in the elaborate science of the magical properties of letters, individually or in groups, a science known as *sīmiyāʾ* or the *ʿilm al-ḥurūf* (MacDonald & Fahd 1997). This science is thought to have begun to develop in the Shīʿah milieu with Jaʿfar al-Ṣādiq (d. AD 765), who is said to have played a major role (Lory 2004: 40). It spread into Sunnī and Ṣūfī contexts in the ninth century AD and, as discussed by Lory, it then went in two different directions: the first being the subject of mystical speculations found in works of mystics such as Ibn ʿArabī (d. AD 1240), and the second which was developed in books about magic the most famous of which is al-Būnī's *Shams al-Maʿārif* (al-Būnī [n.d.]; Lory 1989).

The roots of using magical words and symbols clearly derived from the pre-Islamic era with obvious survivals into Islam through a number of different routes. There is a proliferation of literature and objects from the Near East continuing traditions of Assyrian, Babylonian, or Egyptian magical practices. These are, for example, in the form of amulets written in Aramaic on metal sheets to ward off evil, or to heal, or to gain love, and a large body of incantation texts written on bowls in Jewish Aramaic, Syriac, or Mandaic. There are Nabataean and Jewish amulets and incantation texts which offer close parallels to Islamic examples;[2] magical papyri in Greek and Coptic and much else. In the Islamic era there are therefore evident continuities with past traditions, and

[1] Some aspects of this topic were discussed in Porter 2009.

[2] See for example Naveh 1979: pl. 14. I am grateful to Michael Macdonald for this reference.

Figure 1. *A rock crystal amulet-seal engraved in reverse with strings of letters in Linear Kufic (15 × 7 mm). British Museum 1883 10–31 17. (Image reversed).*

particular elements, such as the so-called "lunette" script (Canaan 1937–1938: 141–143), survive and are placed in a new context. For, in Islam, the amulet becomes a supercharged prayer and the use of letters within the amulets is intimately connected to Muslim beliefs.

That a variety of languages and scripts continued to be used for amulets well into the Islamic period can be seen in the early medieval paper amulets and magic books in the Cairo Geniza which, while their owners spoke Arabic, use a mixture of Hebrew, Aramaic, and Arabic in their formulae (Naveh & Shaked 1985: 30). Canaan, in his comprehensive analysis of amulets, suggested that using words of foreign origin — Hebrew, Syriac, or Greek for example — was rooted in the belief that they were more efficacious than Arabic, and he cited authors such as al-Talmasānī who clearly admitted that this was his reason for using foreign words (Canaan 1937–1938: 91).

Content

Ludvik Kalus was the first to identify a group of the early Arabic magical inscriptions, particularly on amulets, as containing words which might connect them to rain-making ceremonies (Kalus 1981: 93–98; 1987). Other inscriptions contain letter clusters which appear to be very loosely based on the mysterious letters of the Qurʾān (Fig. 1). These letters, which appear in groups immediately before the texts of twenty-nine of the 114 surahs and which were believed to contain particular talismanic powers, make a frequent appearance on amulets (Canaan 1937–1938: 94). A recurring letter string on some examples appears to begin with *kāf-hāʾ-yāʾ-ʿayn-ṣād* from the beginning of *Sūrat Maryam* (19), and continuing with other letters but perhaps using this group of mysterious letters as a starting point (Porter 2009: 144–145). In some cases there is a substitution of *sīn* for *ṣād*.[3] The use of the mysterious letters on amulets is highlighted by an interesting example of an early rock crystal amulet-seal in the National Museum of Iran ascribed to the ninth century AD (Ghouchani 2007: 82) (Fig. 2) where groups of the mysterious letters are included among a number of verses from the Qurʾān (17: 82; 68: 1–2; 21: 87). Other amulet-seals however (as discussed below) contain seemingly random uses of letter clusters which appear to get more abstract over time.

In summary, while some of these objects may contain complex words and phrases connected to rain and water, and others perhaps use the "mysterious letters" as a basis, there are so many variations in the letter clusters that they defy any attempt to categorize them.

The style of the script and where these inscriptions appear

On magic bowls (Fig. 3) from the twelfth century AD and later, the Linear Kufic texts are in cartouches or roundels and in clear contrast to the "legible" inscriptions which are in *naskh* and generally consist of quotations from the Qurʾān or are benedictory in content. This combination of the two scripts continues on magic bowls even until the nineteenth century. On the stone amulets or amulet-seals, Linear Kufic is found on its own where, depending on the object, it appears more or less legible, as already mentioned.

There are three key features to the Linear Kufic style: the shape of the letters, the continuous base line, and its appearance in positive and negative (sometimes both). The shape of the letters is very much associated with early Arabic letter forms. Typical are the forms of the letters *hāʾ*, *ʿayn*, and *ḥaʾ* (Figs 1–2). These may be compared with

[3] Emily Savage-Smith pointed out this substitution. However, it is not clear why it was made. It is possible that it could be associated with the creation of the Eastern *abjad* system where the letter *sīn* replaces the letter *ṣād* as the fifteenth letter with the value of 60. This was suggested to me by Michael Macdonald (personal communication and see 1986: 123–124). For a discussion of the two systems see George 2009: 92.

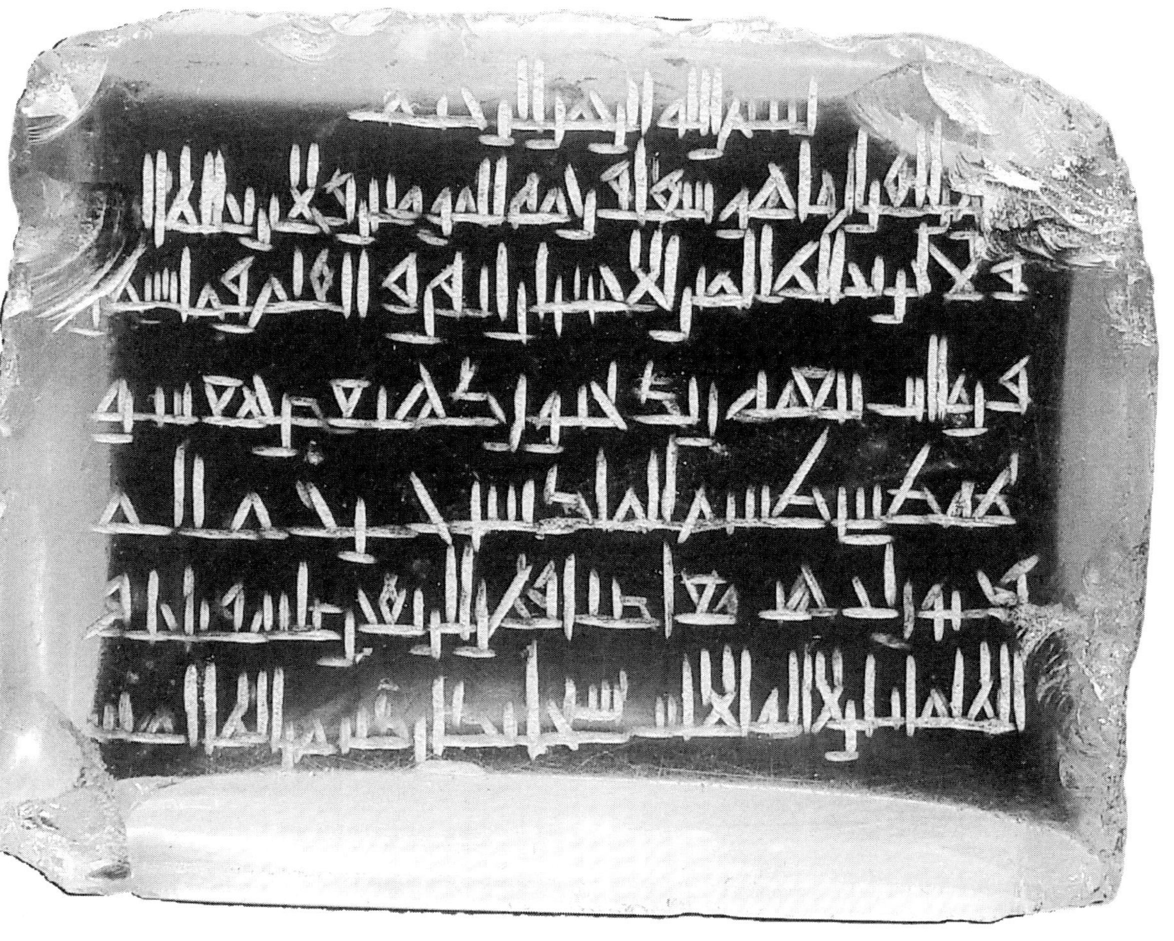

Figure 2. *A rock crystal amulet-seal engraved in reverse with Quranic inscriptions and "mysterious letters" of the Qurʾān. National Museum, Tehran (by permission of Abdullah Ghouchani). (Image reversed).*

early forms highlighted by Grohmann (1971: Schrifttafel II) and which appear in rock inscriptions from the first two centuries of Islam (al-Kilābī 1430/2009: 61–340). However, the problem with such comparisons is that while it may be fairly clear when particular letter forms start to appear, it is difficult to establish precisely how long they continue in use. Care should also be taken when comparing script styles on materials in different media as these may appear and change at different times and for a variety of reasons (Tabbaa 1994).

The writing of the texts in continuous lines, without breaks between individual words (Figs 1–7), is not unique to these magical inscriptions. It appears on glass stamps of the ninth century AD (Morton 1985: 128), and on seal inscriptions (Fig. 6) (Porter, forthcoming). It is also a feature of some Nabataean graffiti.[4] A third feature of the style is that the inscriptions can be written in positive or in reverse, perhaps as a way of further obfuscating the text and rendering it more "magical". An intriguing parallel for the early use of reversed script is one of the Ḥanākiyyah inscriptions in Saudi Arabia datable to the first two centuries of Islam (Donner 1984: W7; Moraekhi 2002: 125) where a prayer for forgiveness is, unusually, written in reverse. On the other hand, the fact that these inscriptions in reverse are on amulets begs the question of how they were used, and suggests that they may have been intended for stamping onto something. However, there is no evidence at the present time of what kind of material

[4] Examples of these can be seen in *CIS* ii nos 1108, 1857, 2236, 2237. I am grateful to Michael Macdonald for this reference.

Figure 3. *Detail of a brass magic bowl showing inscriptions in* naskh *and Linear Kufic (diameter 18.8 cm). British Museum OA+2603.*

Figure 4. *A green jasper amulet-seal engraved in reverse with a string of letters in Linear Kufic (20 × 19 mm). British Museum 1819 7-9 6. (Image reversed).*

Figure 5. *A clay bulla with a lion facing a scorpion in the centre surrounded by stylized strings of Linear Kufic (29 × 29 mm). British Museum 1991 7-27 4.*

Figure 6. *A sardonyx seal inscribed "God is the trust of Salīm" (17 × 12 mm). British Museum 1878 12-20 11. (Image reversed).*

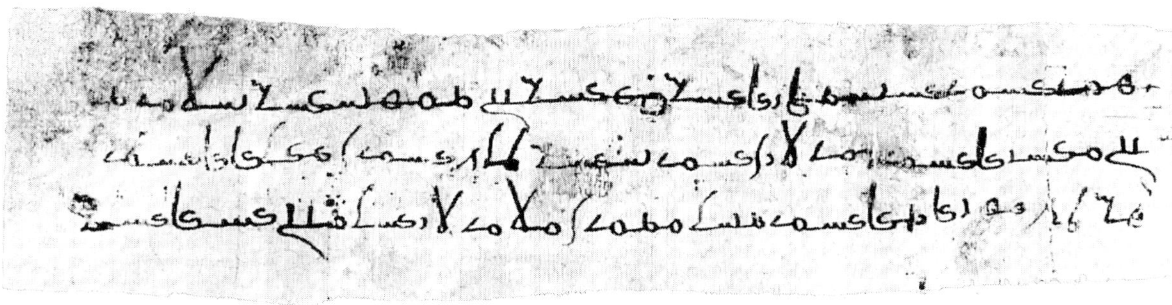

Figure 7. *A papyrus document with "magical" texts. Private collection (by courtesy of Geoffrey Khan).*

this might have been and it may simply be, as Callieri has suggested for ancient Indian seals where this phenomenon also occurs, that these objects which he terms "amulet seals" are in reverse in order to reinforce their mysterious aspect and are not seals as such (2001: 27).

Linear Kufic appears not only on amulets. It is found in Arabic papyri of which two examples can be cited: one is in the Princeton University collections of papyri (CD NS29; Porter 2009: 154), and one in a private collection (Fig. 7).[5] The latter was pointed out to me by Geoffrey Khan who commented that among early Arabic papyri from Egypt the "magical" ones stand out by their use of more angular letter forms in addition to running strings of letters.[6] These magical papyri have been broadly dated to between the eighth and tenth centuries AD since they do not contain anything that at present provides any greater precision.

An intriguing "magical" text painted on pottery from Susa was also pointed out to me recently by Ludvik Kalus.[7] Pottery painted with inscriptions in black ink is fairly common among the unglazed pottery at Susa (Joel & Peli 2005: 94–111) and is dated to the eighth–ninth centuries AD. It is also found among the pottery at the Persian Gulf city of Sīrāf. Some of the Susa examples are inscribed in Aramaic and belong within the context of the Mandaic incantation-bowl tradition, others inscribed in Arabic, where the script corresponds to the more cursive style on the papyri, may be ostraca. What is interesting, therefore, is that the example being studied by Kalus from Susa is in an angular style and with groups of letters written in both directions, thus linking it closely with the "magical" papyri, which are also distinct from the normal papyri. These are at present isolated examples and it is not clear how common these objects were.

Kufic was used up to about the twelfth century AD for writing in official contexts such as coins, epitaphs, etc. and became increasingly elaborate in these contexts as time went on. As far as amulets on paper are concerned, the early block-printed amulets known as *tarsh* are in Kufic with later ones in *naskh*. These were intended to be read. Their texts are generally verses from the Qurʾān; only a very few published examples contain other "magical" elements such as mysterious letters or a five-pointed star (Schaefer 2006: 194). There is no Linear Kufic among the early examples. One example dateable to the thirteenth–fourteenth centuries AD, however, includes among the cursive texts one in a form of Linear Kufic which includes groups of letters and numbers around a central roundel (Bulliet 1987: 437). This use of both script styles on the same object is paralleled on the magic bowls referred to earlier.

In addition to the block-printed amulets were those that were simply handwritten — a tradition which continues — where in among the legible texts appear mysterious and magical words. An example from Quṣayr *c.* AD 1200–1300 (Guo 2004: 311), is a text in which a woman is requesting the birth of a boy. It contains on the second line the letters which combine to make the text *lataḥīṭahṭīl*. The fifteenth-century AD polymath, al-Suyūṭī, in his *Al-raḥmah fī ʾl-ṭibb wa-ʾl-ḥikmah*, describes a series of seven variants of these letters which can be used to grant a woman fertility (Guo 2004: 312). Also from Quṣayr are talismans with magical texts written on cloth (Handley & Regourd 2009: 144).

[5] I am extremely grateful to Geoffrey Khan for drawing this to my attention at the Special Session of the Seminar for Arabian Studies at which this paper was delivered, and for allowing me to illustrate it here. I am also grateful to John Healey with whom I discussed the Princeton papyrus to see if there were any possible Nabataean antecedents in the letter forms.

[6] These comments were made during the discussion following the presentation of the papers at the Special Session.

[7] Personal communication from Ludvik Kalus. This example was shown to him by Monique Kervran and is currently under study. The inscribed sherds from Sīrāf are being studied by Seth Priestman and myself.

Magical alphabet literature

It is worth considering at this point a genre of magical literature that focuses on magical alphabets. The earliest of these texts is attributed to the mysterious Ibn Waḥshiyyah. His book, *Shawq al-mustahām fī maʿrifat rumūz al-aqlām* ("The frenzied devotee's desire to learn about the riddles of ancient scripts") was written in about 241/885 (Ibn Waḥshiyyah 1806). Ibn Waḥshiyyah is himself a mysterious character and it is thought that the real author, possibly compiler or, as some have argued, forger, was Abū Ṭālib al-Zayyāt who claimed to be a pupil or secretary of Ibn Waḥshiyyah (Fahd 1971). Fahd has suggested that the writings of Ibn Waḥshiyyah should be considered "as the result of various successive re-writings and revisions of scientific and pseudo-scientific materials surviving from antiquity, preserved, amplified and modified by Syrian and Alexandrian Hellenism and carried on until the period of the translators of the *Bayt al-ḥikmah*, either by Greek documents or by Pahlavi and Syriac versions" (1971). This was the period of the transmission of ideas and translations of texts where the languages, cultures and religions of the ancient world were still known. Al-Nadīm emphasizes this in the *Fihrist*, completed probably about AD 990, in his descriptions of the different scripts that were known at the time, fanciful or otherwise (al-Nadīm 1970: 9–38).

Ibn Waḥshiyyah's text, translated by Hammer in 1806, is a collection of eighty alphabets and the author lists as his object to "collect the rudiments of alphabets used by ancient nations, doctors and learned philosophers in their books of science for the use of the curious and the studious who apply themselves to philosophical and mystic sciences." The scripts that he enumerates start with Kufic and Maghribī, then include a whole range of scripts — Syriac, Nabataean, Hebrew, *musnad* ("Himyaritic"),[8] then going on to Egyptian hieroglyphs with the alphabets becoming increasingly mystical and esoteric. One, he says, is the alphabet of the twelve constellations, which contains the solution to secrets and the key to treasures (Ibn Waḥshiyya 1806: 15) and is specifically for talismans. A nineteenth-century amulet plaque in the British Museum (Fig. 8) showing Solomon and his jinns has symbols that could have been taken from a compendium of magical alphabets such as Ibn Waḥshiyyah's and demonstrates the continued use of different magical scripts. Around the sides the fanciful letters can be closely paralleled to those of Ibn Waḥshiyyah (1806: 48) (Fig. 9) while in the centre of the plaque are the angular forms of Linear Kufic.

FIGURE 8. *A nineteenth-century brass talismanic plaque with Solomon and his jinns at the top, magical inscriptions below (115 × 90 mm). British Museum OA+ 7408.*

Although Ibn Waḥshiyyah does not refer to Kufic or Maghribī being used in the context of making amulets as he does with some of the other scripts, it is significant that both are included in his compendium, for this is also paralleled in other texts of magical alphabets. Several manuscripts cited by Casanova in his study of magical scripts and Linear Kufic are in the Bibliothèque Nationale in Paris.[9] These include BN 2675 *Kitāb miftāḥ*

[8] Michael Macdonald has pointed out that the letters supposed to be *musnad* bear no relation to the "real" *musnad* letters forms. He also noted that in the alphabetic order *ṭāʾ/ẓāʾ* follow *ʿayn/ghayn* rather than precede them. The tenth-century AD treatise of the Ismāʿīlī *dāʿī*, Jaʿfar b. Manṣūr al-Yaman (Strothmann 1952), includes a section on Ismāʿīlī use of *al-kitābāt al-sirriyyah* ('secret writing'). Jaʿfar al-Yaman describes how this magical script relates to the Yemeni "Himyarite" (Ancient South Arabian). For the continued use on a modern mirror of what was believed to be the Ancient South Arabian script, see Regourd 2007: 146. She suggests that the inscription was believed to consist of names in a "Himyarite" language only understandable by the jinn.

[9] I am grateful to Marie-Geneviève Guesdon for allowing me to study the microfiches of some of these manuscripts and to reproduce folio 90 of BN 2676 on Fig. 10.

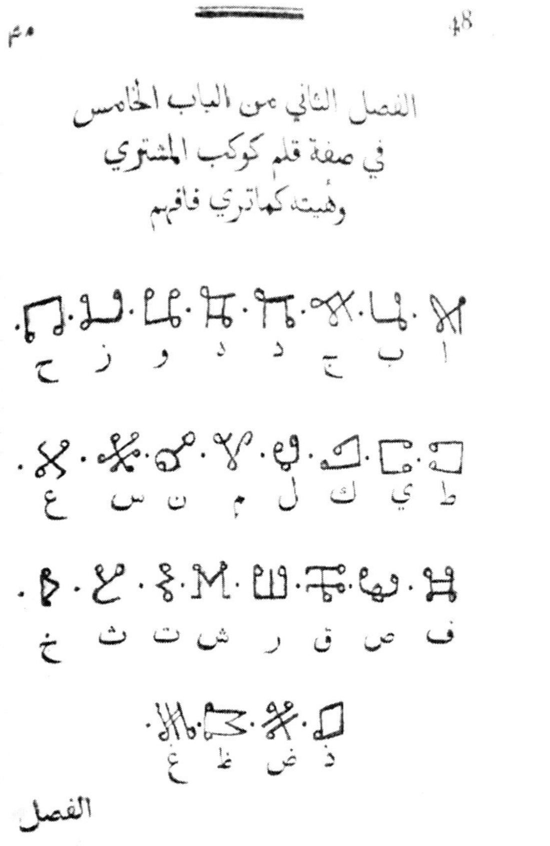

Figure 9. *Description of "the pen of the star of Jupiter". Ibn Waḥshiyyah 1806: 48, detail.*

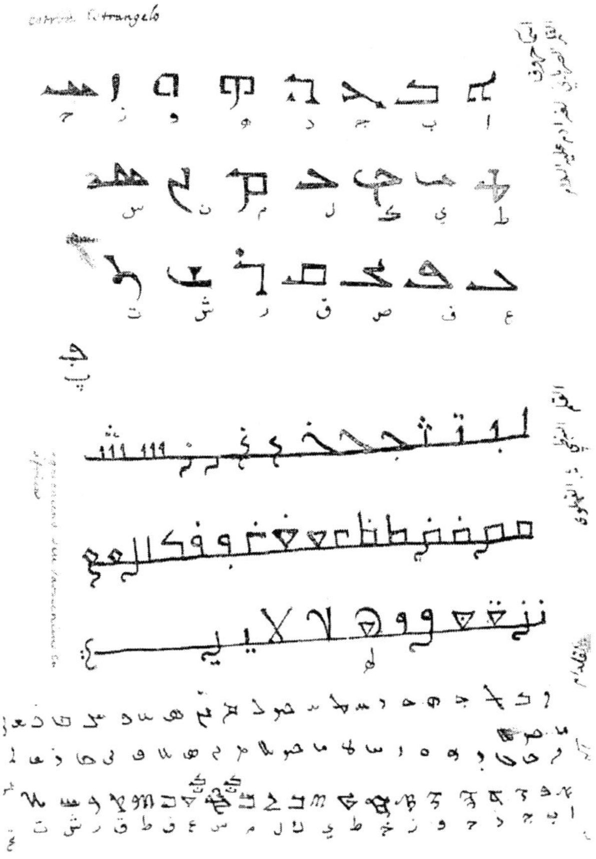

Figure 10. *Folio 90 from BN 2676,* Kitāb miftāḥ asrār ʿulūm al-anbiyāʾ, *showing the Arabic alphabet in the form of continuous letter strings in Linear Kufic. (Courtesy Bibliothèque Nationale).*

asrār ʿulūm al-anbiyāʾ, written in 703/1303–1304, which consists of twenty-four alphabets, many of them imaginary, where there are symbols with Arabic letter equivalents in the *abjad* order of the alphabet.[10] Another is BN 2676, a seventeenth-century copy of the last but with additional scripts added at the end. The interesting point here is that folio 90 of BN 2676 (Fig. 10) has the Arabic alphabet written out on three lines as continuous text resembling the strings of letters on the amulets (see Fig. 1). As can be seen, the order has the letters organized in what Macdonald has called the "eastern formal order", i.e. by shape rather than according to the *abjad* (1986: 123), though curiously, while final *hāʾ* and *ṭāʾ marbūṭah* are in the correct position for *hāʾ*, initial *hāʾ* and *lām-alif* are placed between *wāw* and *yāʾ*. It was this manuscript that led to Casanova's identification of Linear Kufic as a "magical" script. A further manuscript of magical scripts, BN 6805, also includes Kufic and resembles the text of Ibn Waḥshīyyah.

Although there are at present no early manuals showing formulae in Linear Kufic, they must have existed, particularly as, by the twelfth century AD, Linear Kufic letter strings appear in al-Būnī's *Shams al-Maʿārif* (al-Būnī [n.d.]: 244 ff.). This was a text regarded as a synthesis of everything that was known up until this time about the magic of letters, the Names of God, and amulets (Fahd 1987: 232; Lory 1989). The continuity and diffusion of this style of script for magical formulae can be clearly seen by its appearance in the fifteenth-century AD *Libro de Dichos Maravillosos* — a compendium of texts on divination and amulets that includes formulae for various

[10] These include a table of "a lunette" script with its Arabic letter equivalents. I am afraid I did not note whether the *abjad* is in the *maghribī* or *mashriqī* order. See Macdonald 1986: 130.

purposes (Labarta 1993: 12–19; 41). What is interesting in the present context is that the *Libro de Dichos* uses all the elements of the magical vocabulary: lunette script, the "seven signs", magic squares, many of which are in Linear Kufic. They include texts which contain Quranic inscriptions or benedictory phrases written as continuous texts, magical words (1993: 40–41), and strings of letters some of which begin with the letter *kāf* (1993: 46) as do many of the inscriptions on the amulets which, it was argued above, were probably phrases based on the group of mysterious letters *kāf-hāʾ-yāʾ-ʿayn-ṣād*.

We are hampered by not being able to give precise dates to either the amulets that have been cited or the Arabic papyri, and can only continue to refer to them as "early Islamic". However, what is significant in trying to establish how and when angular Kufic began to be commonly used in magic, is that the compendia of magical alphabets from Ibn Waḥshiyyah onwards (which were presumably made for the use of the creators of amulets) included the Kufic and Maghribī scripts. It could be argued in conclusion, therefore, that as Kufic became increasingly elaborate and illegible from about the tenth century AD onwards and as the cursive scripts, *naskh* in particular, began to take over for general use, some of the writers of amulets began to use Kufic in a deliberately obfuscatory way to increase the importance and power of the amulets, and that it was in this way that Kufic entered the realm of magic.

Siglum

CIS *Corpus Inscriptionum Semiticarum*. Pars II. *Inscriptiones Aramaicas continens*. Paris: Imprimerie nationale, 1889–1954.

References

Bulliet R.

1987. Medieval Arabic *Tarsh*: A Forgotten Chapter in the History of Printing. *Journal of the American Oriental Society* 107: 427–438.

al-Būnī, Abū ʾl-ʿAbbās Aḥmad b. ʿAlī./[Editor unknown]

[n.d.]. *Kitāb shams al-maʿārif al-kubrā wa-laṭāʾif al-ʿawārif.* Beirut: al-Maktabah al-thaqāfiyyah.

Callieri P.

2001. In the land of the magi: demons and magic in the everyday life of pre-Islamic Iran. In R. Gyselen (ed.), Démons et merveilles d'Orient. *Res Orientales* 13: 11–35.

Canaan T.

1937–1938. The decipherment of Arabic talismans. *Berytus* 4: 69–111; 5: 141–151. [Reprinted on pages 125–177 in E. Savage-Smith (ed.), *Magic and Divination in Early Islam.* (The Formation of the Classical Islamic World, 42). Aldershot: Ashgate Variorum, 2004].

Casanova M.

1921. Alphabets magiques arabes. *Journal Asiatique* 11th series 18: 37–55.

Donner F.M.

1984. Some Early Arabic Inscriptions from al-Ḥanākiyya, Saudi Arabia. *Journal of Near Eastern Studies* 43: 181–208.

Fahd T.

1971. Ibn Waḥshiyyah. Pages 963–965 in *Encyclopaedia of Islam.* (New edition). iii. Leiden: Brill.

1987. *La divination arabe.* Paris: Sindbad.

George A.

2009. Calligraphy, Color and Light in the Blue Qurʾan. *Journal of Qurʾanic Studies* 11: 75–125.

Ghouchani A.

2007. Qadimitarin muhr-i nuveshta-dar duran islami az yak kakim-i sasani. *Majalla-ya bastanshenasi va tarikh* (*Iranian Journal of Archaeology and History*) 20/1–2 (39–40): 82–99.

Grohmann A.

1971. *Arabische Paläographie.* ii. *Das Schriftwesen. Die Lapidarschrift.* (Österreichische Akademie der Wissenschaften, Philosophisch-historische Klasse, Denkschriften 94/2). Vienna: Böhlaus.

Guo L.

2004. *Commerce, Culture and Community in a Red Sea Port in the Thirteenth Century: The Arabic documents from Quseir*. Leiden: Brill.

Handley F.J.L. & Regourd A.

2009. Textiles with writing from Quṣeir al-Qadīm. Finds from the Southampton excavations 1999–2003. Pages 141–153 in L. Blue, J. Cooper, R. Thomas & J. Whiteright, *Connected Hinterlands. Proceedings of Red Sea Project IV*. (British Archaeological Reports, International Series, 2052). Oxford: Archaeopress.

Ibn Waḥshiyyah, Aḥmad ibn ᶜAlī/trans. J. Hammer[-Purgstall]

1806. *Shawq al-mustahām fī maᶜrifat rumūz al-aqlām. Ancient alphabets and hieroglyphic characters explained: with an account of the Egyptian priests, their classes, initiation, and sacrifices*. London: Bulmer.

Joel G. & Peli A.

2005. *Suse: terres cuites islamiques*. Ghent: Snoek.

Kalus L.

1981. *Catalogue des cachets, bulles et talismans islamiques*. Paris: Bibliothèque nationale, Département des monnaies, médailles et antiquités.

1987. Rock Crystal Talismans Against Drought. Pages 101–105 in N. Brosh & M. Rosen-Ayalon (eds), *Jewellery and goldsmithing in the Islamic World*. Jerusalem: Israel Museum.

al-Kilābī Ḥ.ᶜA.

1430/2009. *Al-nuqūsh al-islāmiyyah ᶜalā ṭarīq al-ḥajj al-shāmī bi-shimāl gharb al-mamlakah al-ᶜarabiyyah al-suᶜūdiyyah*. Riyadh: Maktabat al-Malik Fahd.

Labarta A.

1993. *Libro de Dichos Maravillosos*. (Misceláneo Morisco de magia y adivinación). (Fuentes Arábico-Hispanas, 12). Madrid: Consejo Superior de investigaciones scientíficas instituto de cooperación con el mundo Árabe.

Lory P.

1989. La magie des lettres dans le Šams al-Maᶜārif d'al-Būnī. *Bulletin des Études Orientales* 39–40: 97–111.

2004. *La science des lettres en terre d'Islam*. Paris: Dervy.

MacDonald D.B. & Fahd T.

1997. Sīmiyāʾ. Pages 612–613 in *Encyclopaedia of Islam.* (New edition). ix. Leiden: Brill.

Macdonald M.C.A.

1986. ABCs and letter order in Ancient North Arabian. *Proceedings of the Seminar for Arabian Studies* 16: 101–168.

Moraekhi M.

2002. A New Perspective on the Phenomenon of Mirror-Image Writing in Arabic Calligraphy. Pages 123–135 in J.F. Healey & V. Porter (eds), *Studies on Arabia in Honour of Professor G. Rex Smith*. (Journal of Semitic Studies Supplement, 14). Oxford: Oxford University Press.

Morton A.H.

1985. *A Catalogue of Early Glass Stamps in the British Museum*. London: British Museum Press.

al-Nadīm, Muḥammad ibn Isḥāq/ed. and trans. B. Dodge

1970. *The Fihrist of al-Nadīm. A tenth-century survey of Muslim culture*. (Records of Civilization: Sources and Studies, 83). New York: Columbia University Press.

Naveh J.

1979. A Nabatean Incantation Text. *Israel Exploration Journal* 29: 111–119.

Naveh J. & Shaked S.

1985. *Amulets and magic bowls: Aramaic incantations of late antiquity*. Brill: Leiden.

Porter V.

2009. Stones to bring Rain? Pages 131–151 in S.S. Blair and J. Bloom (eds), *Rivers of Paradise: water in Islamic art and culture*. New Haven, CT: Yale University Press.

(forthcoming). *Catalogue of Arabic and Persian Seals and Amulets in the British Museum.* London: British Museum Press.

Regourd A.

2007. A magic mirror in the Louvre and additional observations on the use of magic mirrors in contemporary Yemen. Pages 135–155 in F. Suleman (ed.), *Word of God Art of Man. The Qurʾan and its Creative Expressions.* Oxford: Oxford University Press/Institute for Ismaili Studies.

Savage-Smith E.

1997. *Science Tools and Magic.* (The Nasser D. Khalili Collection of Islamic Art, 12, parts 1 and 2). London: Azimuth Editions/Oxford University Press.

Schaefer K.R.

2006. Enigmatic Charms: Medieval Arabic Block Printed Amulets in American and European Libraries and Museums. (Handbook of Oriental Studies, 1/82). Leiden: Brill.

Strothmann R.

1952. *Kitāb al-kashf of Jaʿfar b. Manṣūr al-Yaman.* London: Oxford University Press.

Tabbaa Y.

1994. The Transformation of Arabic Writing: Part 2, the Public Text. *Ars Orientalis* 24: 119–147.

Author's address

Venetia Porter, Department of the Middle East, The British Museum, London WC1B 3DG, UK.

e-mail vporter@thebritishmuseum.ac.uk

M.C.A. Macdonald (ed.), *The development of Arabic as a written language.* (Supplement to the Proceedings of the Seminar for Arabian Studies 40). Oxford: Archaeopress, 2010, pp. 141–143.

The Old Arabic graffito at Jabal Usays: A new reading of line 1

M.C.A. MACDONALD

FIGURE 1. *The Old Arabic graffito at Jabal Usays. (Photograph M.C.A. Macdonald, 2006).*

The history of the study of the Old Arabic graffito at Jabal Usays[1] is outlined by Pierre Larcher in this volume. Here, I would like to present a new reading of the first line which has been made possible by a photograph I took at the site in November 2006.[2] I have published a note on this elsewhere (2009).[3] However, since the inscription is mentioned several times in the present volume, it seemed sensible to include, for the convenience of the reader, a discussion of my new reading, together with the photograph (Fig. 1).

In 2002, Robin and Gorca read line 1 as 'nh(.) *Qṯm* bn Mġ(y)r(h) *ʾl- ʾwsy.*[4] He later changed this to "*Je suis* Ruqaym, *fils de* Muʿriḍ *l'Awsite*".[5] Independently, I had reached the same reading of the first name, i.e. *rqym*. The *r* is clear on the new photograph, as it is in the next word, *br*, showing that the traditional reading of this word is correct, as opposed to Robin's suggestion to read *bn* here and in the other pre-Islamic inscriptions in the Arabic script (Robin & Gorea 2006: 508). Compare the *r* of *rqym* and *br* with those in *ʾrslny* and *ʾl-ḥrṯ* in line 2.

Aramaic *br* for Arabic *bn* also occurs in the Namārah inscription (AD 328), which is otherwise in the Old Arabic language expressed in the Nabataean Aramaic script, and is found together with several other Aramaic words and phrases (*šlm*, *dkyr*, *b-ṭb*, etc.) in graffiti in the

[1] *Usays* is the ancient name as recorded in Safaitic inscriptions (Macdonald Al Muʾazzin & Nehmé 1996: 466, and in the Old Arabic graffito as shown by the new reading published by Robin & Gorea 2002: 507, 509). *Jabal Says* is the modern name give in the *Official Standard Names Gazetteer* for Syria (Washington, DC, 1967) and on maps, Google Earth, etc.

[2] Although I made this new reading available to Pierre Larcher before the Special Session in July 2009, he preferred to retain those arguments in his paper which are based on previous readings (see Larcher, this volume: 106, n. 2).

[3] The photograph mentioned in that note is on p. 120 of *Semitica et Classica* 2.

[4] Robin & Gorea 2002: 507; Robin 2006: 331, (uncertain readings in roman type). This reading was based on Robert Hoyland's photograph taken in 2001.

[5] Robin 2008: 178. Only the translation was published, with uncertain readings in roman type. This was based on my photograph (see Fig. 1).

Nabataean script in Sinai and north-west Arabia. In some of the latter, Arabic words, phrases, and/or syntax occur, suggesting that they were composed by Arabic-speakers (see, for instance, Nehmé, this volume: 69 JSNab 18, 71 S 1, 76 UJadh 109). These Aramaic words seem to have been used roughly in the same way as we employ such Latin expressions as *pace*, *caveat emptor*, *sine qua non*, or the way that the Latin vocative and ablative in the word *Jesu* are preserved in some English prayers of the Anglican church, etc. Such a usage does not "mettre en doute la cohérence linguistique" of the texts in which they occur (Robin & Gorea 2002: 508), but is simply a relic of a learned or written language which is being, or has been, displaced as the spoken one is increasingly used for writing.

In the second name, the first three letters are clearly *m-ᶜ/ġ-r/z*. The last letter is clearly a *f*, and cannot be a *ḍ* as suggested by Robin (2008: 178). The tail of final *ṣ/ḍ* in both Nabataean and early Arabic always runs downwards, not horizontally/diagonally, and the top of the original stem is always present as a short vertical line immediately to the left of the loop, as it still is in modern forms of the Arabic script.[6]

By contrast, from at least the second century AD in scribal hands and perhaps a century later in inscriptions, the *final* form of *f*, has virtually no stem and is simply a loop with a long horizontal/diagonal tail, even though its initial/medial form often retains the original vertical stem, which, however, never rises above the loop as it does in *ṣ/ḍ*. Compare the shape here (AD 528/529) with that in the name *ywsp* at the end of line 2 in the Nabataean inscription from Taymāʾ dated to AD 203 (al-Najem & Macdonald 2009: 209), in the same name in the "transitional" texts in Nehmé, this volume: Fig. 27 line 2, Fig. 41 line 1, and in the word *yastankif* in the north and north-east sections of the mosaic inscription in the inner octagonal arcade of the Dome of the Rock (AD 692). The second name must therefore read *m-ᶜ/ġ-r/z-f* and of these possibilities only *mᶜrf* would seem to produce a recognizable name: *Muᶜarrif.* I am most grateful to Pierre Larcher (personal communication) for telling me that this name is used today and to Robert Hoyland (personal communication) for the information that, though uncommon, it is known from texts of the early Islamic period.[7] Thus the first line would now read:

ʾnh rqym br mᶜrf ʾl-ʾwsy

"I, Ruqaym son of Muᶜarrif the Awsite...."

[6] For Nabataean, see the script table on Macdonald 2003: 53, fig, 38, where in line 15 the final forms of *ṣ/ḍ* and *p/f* in a Nabataean scribal hand can be compared. For early Arabic, see Gruendler 1993: 69–71.

[7] See, for instance, Muᶜarrif ibn Wāṣil in Ibn Saᶜd 1957–1960: 1.389, 5.403, 6.45, 100, 356.

References

Gruendler B.

1993. *The Development of the Arabic Scripts. From the Nabatean Era to the First Islamic Century According to Dated Texts.* (Harvard Semitic Studies, 43). Atlanta, GA: Scholars Press.

Ibn Saᶜd, Abū ᶜAbd Allāh Muḥammad/[Editor unknown]

1957–1960. *Al-Ṭabaqāt al-kubrā.* Beirut: Dār Ṣāder.

Larcher P.

(this volume). In search of a standard. Dialect variation and New Arabic features in the oldest Arabic written documents. Pages 102–112 in M.C.A. Macdonald (ed.), *The development of Arabic as a written language.* (Supplement to Proceedings of the Seminar for Arabian Studies 40). Oxford: Archaeopress.

Macdonald M.C.A.

2003. Languages, scripts and the use of writing among the Nabataeans. Pages 36–56, 264–266 (endnotes), 274–282 (references) in G. Markoe (ed.), *Petra Rediscovered: lost city of the Nabataeans.* New York: Abrams/Cincinnati, OH: Cincinnati Art Museum.

2009. A note on New Readings in Line 1 of the Old Arabic Graffito at Jabal Says. *Semitica et Classica* 2: 223.

Macdonald M.C.A., Al Muᵉazzin M. & Nehmé L.

2006. Les inscriptions safaïtiques de Syrie, cent quarante ans aprés leur découverte. *Comptes rendus de l'Académie des Inscriptions & Belles-Lettres*: 435–494.

al-Najem M. & Macdonald M.C.A.

2009. A new Nabataean inscription from Taymāʾ. *Arabian Archaeology and Epigraphy* 20: 208–217.

Nehmé L.

(this volume). A glimpse of the development of the Nabataean script into Arabic based on old and new epigraphic material. Pages 47–88 in M.C.A. Macdonald (ed.), *The development of Arabic as a written language.* (Supplement to Proceedings of the Seminar for Arabian Studies 40). Oxford: Archaeopress.

Robin C.J.

2006. La réforme de l'écriture arabe à l'époque du califat médinois. *Mélanges de l'Université Saint-Joseph* 59: 319–364.

2008. Les Arabes de Ḥimyar, des « Romains » et des Perses (IIIe–VIe siècles de l'ère chrétienne). *Semitica et Classica* 1:167–207.

Robin C.J. & Gorea M.

2002. Un réexamen de l'inscription arabe préislamique du Ǧabal Usays. *Arabica* 49: 503–510.

Author's address

Michael C.A. Macdonald, Oriental Institute, Pusey Lane, Oxford, OX1 2LE

e-mail michael.macdonald@orinst.ox.ac.uk.

Papers read in the Special Session of the
Seminar for Arabian Studies on 24 July 2010

Papers included in this volume are marked with a *

*Christian Robin	Introduction
*Michael Macdonald	*Why did Arabic remain a purely spoken language for so long?* [Incorporated in *Ancient Arabia and the written word* published here]
*Laïla Nehmé and *ᶜAlī Al-Ghabbān	*New very late Nabataean, "transitional", and early Arabic inscriptions, with a comparison of their content.*
*Robert Hoyland	*Power, Patronage and Arabic inscriptions.*
*Pierre Larcher	*In search of a standard: dialect choices in the development of Classical Arabic.*
*François Déroche	*The Codex Parisino-petropolitanus and the* Ḥijāzī *scripts.*
Alain George	*On the roots and context of the* Ḥijāzī *corpus.*
Marcus Fraser	*Qurᵓans in "*Ḥijāzī*" scripts: marshalling the evidence.*
*Venetia Porter	*The use of writing in magic.*
*Gregor Schoeler	*The relationship of literacy and memory in the second/eighth century.*

This was followed by a workshop in which discussion of the papers continued, and particular inscriptions, manuscripts, documents, coins and objects of relevance to the subject of the Special Session were presented.

Index

Note, diacritical marks are ignored in the alphabetic order

Epigraphic Index

The works cited in the left column can be found in the lists of references in the papers the page numbers of which appear in the right column.

Inscriptions

Aramaic

Aramaic script of Taymāʾ

CIS ii 336

Christian Palestinian Aramaic

Those texts discussed in this volume are listed here. Those listed by Robert Hoyland but not discussed in his paper will be found in his Appendix 1 on pp. 37 and 39.

Imperial Aramaic

Nabataean

For the purposes of this index the distinctions between "classical" Nabataean and "transitional" inscriptions are ignored.

Manuscripts

Ostraca

Papyri

References to the Bible

References to the Qurʾān